Mathematical Connections

A Companion for Teachers and Others

Mathematical Connections was developed at Education Development Center, Inc. (EDC) within the Center for Mathematics Education (CME) with partial support from the National Science Foundation.

This material is based on work supported by the National Science Foundation under Grant No. ESI-9617369. Any opinions, findings and conclusions or recommendations expressed in this material are those of the author(s) and do not necessarily reflect the views of the National Science Foundation.

ISBN 0-88385-739-1

Library of Congress Catalog Control Number 200592349

Current Printing (last digit):
10 9 8 7 6 5 4 3 2 1

Mathematical Connections

A Companion for Teachers and Others

Al Cuoco
Center for Mathematics Education
Education Development Center, Inc.

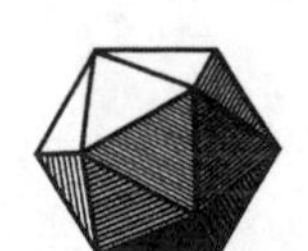

Published and Distributed by
THE MATHEMATICAL ASSOCIATION OF AMERICA

CLASSROOM RESOURCE MATERIALS

Classroom Resource Materials is intended to provide supplementary classroom material for students—laboratory exercises, projects, historical information, textbooks with unusual approaches for presenting mathematical ideas, career information, etc.

101 Careers in Mathematics, 2nd edition, edited by Andrew Sterrett
Archimedes: What Did He Do Besides Cry Eureka?, Sherman Stein
Calculus Mysteries and Thrillers, R. Grant Woods
Combinatorics: A Problem Oriented Approach, Daniel A. Marcus
Conjecture and Proof, Miklós Laczkovich
A Course in Mathematical Modeling, Douglas Mooney and Randall Swift
Cryptological Mathematics, Robert Edward Lewand
Elementary Mathematical Models, Dan Kalman
Environmental Mathematics in the Classroom, edited by B. A. Fusaro and P. C. Kenschaft
Essentials of Mathematics: Introduction to Theory, Proof, and the Professional Culture, Margie Hale
Exploratory Examples for Real Analysis, Joanne E. Snow and Kirk E. Weller
Fourier Series, Rajendra Bhatia
Geometry from Africa: Mathematical and Educational Explorations, Paulus Gerdes
Identification Numbers and Check Digit Schemes, Joseph Kirtland
Interdisciplinary Lively Application Projects, edited by Chris Arney
Inverse Problems: Activities for Undergraduates, C. W. Groetsch
Laboratory Experiences in Group Theory, Ellen Maycock Parker
Learn from the Masters, Frank Swetz, John Fauvel, Otto Bekken, Bengt Johansson, and Victor Katz
Math through the Ages: A Gentle History for Teachers and Others (Expanded Edition), William P. Berlinghoff and Fernando Q. Gouvêa
Mathematical Connections: A Companion for Teachers and Others, Al Cuoco
Mathematical Evolutions, edited by Abe Shenitzer and John Stillwell
Mathematical Modeling in the Environment, Charles Hadlock
Mathematics for Business Decisions Part 1: Probability and Simulation (electronic textbook), Richard B. Thompson and Christopher G. Lamoureux
Mathematics for Business Decisions Part 2: Calculus and Optimization (electronic textbook), Richard B. Thompson and Christopher G. Lamoureux
Ordinary Differential Equations: A Brief Eclectic Tour, David A. Sánchez
Oval Track and Other Permutation Puzzles, John O. Kiltinen
A Primer of Abstract Mathematics, Robert B. Ash
Proofs Without Words, Roger B. Nelsen
Proofs Without Words II, Roger B. Nelsen
A Radical Approach to Real Analysis, David M. Bressoud
Resources for the Study of Real Analysis, Robert L. Brabenec
She Does Math!, edited by Marla Parker
Solve This: Math Activities for Students and Clubs, James S. Tanton
Student Manual for Mathematics for Business Decisions Part 1: Probability and Simulation, David Williamson, Marilou Mendel, Julie Tarr, and Deborah Yoklic

Student Manual for Mathematics for Business Decisions Part 2: Calculus and Optimization, David Williamson, Marilou Mendel, Julie Tarr, and Deborah Yoklic
Teaching Statistics Using Baseball, Jim Albert
Understanding our Quantitative World, Janet Anderson and Todd Swanson
Writing Projects for Mathematics Courses: Crushed Clowns, Cars, and Coffee to Go, Annalisa Crannell, Gavin LaRose, Thomas Ratliff, and Elyn Rykken
Mathematical Connections: A Companion for Teachers and Others, Al Cuoco

MAA Service Center
P. O. Box 91112
Washington, DC 20090-1112
1-800-331-1MAA FAX: 1-301-206-9789
www.maa.org

Dedicated to the Memory of
Johnny Cuoco
1916–2003
Father, friend, and first baseman extraordinaire

Contents

Introduction

This book grew out of our work at Education Development Center on a course for high school seniors as part of a curriculum called *The CME Project* [15]. That senior course provides students who have completed the equivalent of second-year high school algebra an exposure to some important ideas in classical and modern mathematics. More importantly, the course shows students some of the ways in which mathematicians work.

Many curricula concentrate on important results in mathematics. *The CME Project* concentrates on important methods for obtaining those results.

In developing *The CME Project*, my colleagues and I consulted regularly with high school teachers. Almost all of them told us that much of mathematics in the later years of the program was either new to them or buried deep in their undergraduate backgrounds. Together with these teachers, we came up with the idea for this book: a development of some mathematical topics and ideas, written specifically for practicing or prospective high school teachers. There are several things that this book is not:

About the sidenotes: Because this book emphasizes the thinking behind the results, the development is non-linear. Sidenotes give me a (less than perfect) way to evolve several ideas at once. You'll develop your own style for dealing with them.

- This is not a teaching guide for *The CME Project*. That curriculum already contains extensive solutions and teaching notes for the day-to-day teaching of the course.
- This is not a typical mathematics text. The focus on mathematical *habits of mind* leads to a different organization than the definition-theorem-proof-example style of writing. Instead, you'll be let in on some of the ideas that lead up to the results, on different ways to think about topics, and on the connections among the topics. This is more about how mathematics is conceived than how it is presented. More about this approach in a minute.
- This is not a text for students. While many of the topics in the book might be appropriate for your classes, the translations are left to you.
- This is not a methods text. While the book is written with teachers in mind, the focus is on mathematics, not on pedagogy. If you are a teacher, *you* are the

Mathematics has a tradition of presenting results in a logical and deductive style that shows how every fact can be deduced from previous ones. This is an extremely efficient, and often elegant, way of presenting one's results. What's missing from these presentations are the long periods of work that lead up to the final polished products. That's what *this* book is about.

expert on how to get ideas across to your students. You'll find no discussions here about group work, alternative assessment, classroom uses of technology, or classroom dialogue. This book is for you as a mathematician.

What, then, *is* this book about?

It is about a closely knit collection of topics that are at the intersection of algebra, arithmetic, combinatorics, geometry, and calculus. It's about some of the mathematics at the base of modern programs like *The CME Project* that mix discrete and continuous mathematics. It's about some classical mathematics that has become more important (and tractable) because of advances in computational technology. And, most importantly, it is about some mathematical ways of thinking that I've found extremely useful in my high school teaching, both in my role as a mathematician and as a mentor for young people learning to do mathematics. I've picked out a suite of ideas that are a joy to my mind (and that I think *you'll* enjoy), that will be useful in your teaching, that may be new to you, and that brings coherence to some of the topics you teach or will teach.

As a result, this is a rather personal book. You're getting my choice of topics and a look at how a small circle of colleagues thinks about them. There are two implications:

- Other authors might have chosen other topics. I've become convinced, largely by working with other teachers over the past decade, of the value of *depth* (as opposed to breadth) in courses and books for teachers. Surveys have their place, but I wanted to write a book that takes *a few simple themes*—like fitting functions to data—and pushes them for all their worth.
- Other authors might have chosen other approaches. I make no apology for the fact that there's a distinct bias towards algebra and algebraic thinking in most of the chapters. I've found it productive to focus on one thing at a time, and the particular approaches used in this book—things like abstracting regularity from repeated calculations—have been especially helpful in my own work and in my teaching.

The many contributors to the book have had high school mathematics teaching experience, and that, too, has guided the choice of topics and approach. For example, knowing about combinatorial proofs or difference calculus helps me design rich problem sets, field questions in class, connect ideas from one course to another, and know when a student's idea is likely to grow into something that will be valuable for an entire class. But most mathematics courses in combinatorics, calculus, discrete mathematics, or number theory aren't set up to develop applications to the teaching of high school mathematics. It's not that they are bad courses—they have other purposes. As a result, most of what I know about these applications I've had to figure out for myself, learn from other teachers, or come upon by accident in books or articles. Many of my colleagues have had similar experiences and have tricks of the trade of their own. I've borrowed heavily from this collective wisdom. The book was designed for several settings:

- mathematics courses for prospective teachers,
- inservice courses for practicing teachers,

- teacher study groups in a high school department,
- self study.

These purposes determined the book's organization. Each chapter takes up one or two topics and develops them in depth. The heart of each chapter is the problem sets; as you probably tell your students, working the problems is where the fun is.

To facilitate self-study, the problem sets are rather orchestrated: *they are meant to be done in sequence*, and many of them refer to previous problems. In some cases, subsets of problems in a given set are chunked so that you know which problems "go together." Many of the problems ask you to establish results that will be essential later in the book. If you skip any of these "essential" problems, you can go back to them when they are cited in later discussions. You'll also find in the problem sets applications and extensions of the results in the chapter, previews of coming attractions, and connections to topics you teach.

At several points in the text, there are historical notes and references. A very good source for further historical readings is [3].

For too long, professional development for high school teachers has meant either taking yet another university mathematics course (with no connection to what we teach) or taking a workshop about teaching (with no mathematics in it at all). But the foundations of high school mathematics go deep enough to involve sophisticated thinking, hard problems, and subtle connections, all while staying connected to the kind of mathematics we talk about with our students every day. What I've attempted here is to outline one of many possible trips into these foundations of *our* subject. Enjoy.

Some of the problems are hard, and some ask you to fill in holes in the exposition. Many are accompanied by hints or partial solutions at the end of each chapter.

Some sets of problems ask you to do extensive algebraic calculations, not just for the fun of it, but because there's a point to the end results. If you get tired, take a break or skip ahead. If you need something later, there will be pointers to the relevant problems.

Acknowledgements

The writing of this book was supported in part by NSF grant ESI 9617369. So many people contributed so many ideas that it's impossible to list them all. But I have to mention a few:

- Michelle Manes, when she was my colleague at EDC, helped outline, critique, and edit earlier drafts of the book, and she did a significant part of the writing in Chapter 3. Her influence is everywhere.
- Ryota Matsuura, then a high school teacher in Brookline Massachusetts, worked through all the problems and wrote the hints and solutions that accompany each chapter. In addition, he found and helped fix mistakes, suggested improvements, and made everything much better.
- Wayne Harvey, director of our division at EDC, worked in detail through every problem in earlier drafts of Chapter 1, making specific suggestions for improving clarity and structure and pushing me to think about my goals for the book.

I'd also like to thank Steve Maurer, Peter Renz, and Barbara Hubbard for their very thoughtful reviews, Bowen Kerins and Ben Sinwell for trying some of the chapters in their course for teachers at the Park City Mathematics Institute, and

At the time of this writing (January, 2004), both Michelle and Ryota are graduate students in mathematics.

The mistakes that remain are solely my responsibility. I'd very much appreciate hearing about them.

the thousands of high school students and teachers with whom I've been privileged to work over the past three decades.

Beverly Ruedi, Elaine Pedreira, and Don Albers from MAA were very generous with their vast expertise about publishing, design, and production. And my colleagues Helen Lebowitz and Nancy D'Amato at EDC pitched in at several important moments, creating graphics, transferring files, and staying in contact with MAA. All of these wonderful people make it all look easy.

Finally, none of this work (or anything else of value in my life) would have been possible without the love and support of my wife Micky.

— February 1, 2004

Center for Mathematics Education—EDC,
Newton Massachusetts.
`alcuoco@edc.org`

Annotated Table of Contents

Chapter 1. Difference Tables and Polynomial Fits

Given a table like this:

Input	Output
0	0
1	1
2	4
3	10
4	20
5	35

How do you find a simple (polynomial or exponential, say) function that agrees with the table? You may know that the old-fashioned theory of successive differences can be used:

Here, Δ means difference. So, $6 = 10 - 4$, $5 = 15 - 10$, and so on. And Δ^2 means "second difference:" the difference of the difference column.

Input	Output		Δ		Δ^2		Δ^3
0	0						
		$\rangle\rightarrow$	1				
1	1			$\rangle\rightarrow$	2		
		$\rangle\rightarrow$	3			$\rangle\rightarrow$	1
2	4			$\rangle\rightarrow$	3		
		$\rangle\rightarrow$	6			$\rangle\rightarrow$	1
3	10			$\rangle\rightarrow$	4		
		$\rangle\rightarrow$	10			$\rangle\rightarrow$	1
4	20			$\rangle\rightarrow$	5		
		$\rangle\rightarrow$	15			$\rangle\rightarrow$	1
5	35			$\rangle\rightarrow$	6		
		$\rangle\rightarrow$	21				

The third differences are constant, so there's a cubic polynomial that will agree with this data. This fact, without proof, is about as far as any current high school curriculum takes the subject.

In this chapter, we'll develop an extensive theory of finite differences:

And more generally. we'll see why a constant nth difference means an nth degree polynomial will work.

- We'll look at why a constant third difference implies a cubic fit.
- We'll develop several different ways to *find* a polynomial that agrees with such a table. Some of these methods go back to Newton and Lagrange.
- We'll come up with a way to "fool" the table: classifying all polynomials that agree with a given set of points.

This is a standardized test maker's nightmare: there is a unique polynomial of minimal degree that agrees with a table, but there are infinitely many polynomials of higher degree that work, too. We'll see how to get them.

The methods we develop in chapter 1 will be essential throughout the book.

Chapter 2. Form and Function

This is a chapter about algebra. In precollege algebra, we're sort of cavalier about the nature of polynomials and algebraic expressions. When we work with graphs and equations—and when we use a graphing calculator—we're often thinking of polynomials as *functions* that describe variation or that produce outputs from inputs. When we go into "algebra mode" and discuss factoring, simplifying, and expanding—and when we use a CAS—we often treat polynomials as *formal expressions*, elements of an algebraic system that has an internal logic and arithmetic of its own. This chapter is about the connections between the two different ways to think about polynomials, and it's also about some classical algebra that's very useful in teaching but has been lost from most undergraduate courses. We'll see how some results of calculus, like the rule for taking the derivative of a polynomial, arise in a completely algebraic context and are forced on us by the rules of elementary arithmetic. In another direction, we'll generalize the "sum and product of the roots" topic from second-year algebra to more general relations between the roots and coefficients of a polynomial. These relations can be used to put the quadratic formula and the formula for a cubic equation into a more general setting.

CAS is short for "Computer Algebra System." A CAS is a software environment that allows you to simplify, expand, and transform algebraic expressions. Examples are *Mathematica* and the CAS on the TI-89 calculator.

Chapter 3. Complex Numbers, Complex Maps, and Trigonometry

We'll begin with an historical introduction to complex numbers, tracing their origin to the problem of finding roots of cubic equations. Then we'll look at the representation of complex numbers as points on the plane, and we'll give geometric interpretations for addition and multiplication. Our approach to the "multiplication rule" is quite simple. Most treatments use the addition formulas for sine and cosine. But in the summer of 2002, a group of high school teachers at PCMI (the Park City Mathematics Institute in Utah) discovered a proof of the rule that uses only elementary geometry. We'll look at their proof in this chapter. The bonus of this approach is that one can turn the typical development on its head and use the multiplication rule to *derive* the addition formulas for sine and cosine. We'll do that, too.

The *multiplication rule*: to multiply two complex numbers represented as vectors on the complex plane, multiply their lengths and add their arguments. If this sounds foreign to you, don't worry—we'll start from scratch in the chapter.

There's a great debate these days about the role of trigonometric identities in high school mathematics. Are "identities" a topic that can safely be dropped from the curriculum? Is there any rhyme or reason to the pages of exercises that we (sometimes) ask students to do? Are there any really *important* identities for later work in science or mathematics? And how does technology impact what we can or should teach?

In this chapter, we take one path through this complex discussion. Two very important identities are the Pythagorean identity

$$\sin^2 x + \cos^2 x = 1$$

and the addition formulas

$$\sin(x + y) = \sin x \cos y + \cos x \sin y,$$
$$\cos(x + y) = \cos x \cos y - \sin x \sin y.$$

The addition formulas lead to the double angle formulas:

$$\sin(2x) = 2 \sin x \cos x,$$
$$\cos(2x) = 2 \cos^2 x - 1.$$

We'll look at this small important collection and find similar formulas for $\sin nx$ and $\cos nx$ for any positive integer n. This will involve creating a sequence of polynomials that shows up all over mathematics and science. This sequence brings coherence to trigonometry, and it connects trigonometry with arithmetic, algebra, and geometry.

The polynomials are called *Chebyshev polynomials.* Using a CAS, we'll be able to find many beautiful patterns in the sequence of Chebyshev polynomials, patterns that can be proven by algebraic techniques and that have implications for trigonometry and geometry.

We'll next look at a key topic in many secondary programs: visualizing functions. You have several ways to visualize the function $x \mapsto x^2$ when x runs over the real numbers. One of the most useful is a Cartesian graph of the function. But what if x is allowed to run over the complex numbers? How can you picture what's going on? You can visualize the complex numbers as points on a plane. So, to get a "graph" of the function in the usual sense, you'd need four dimensions, and we don't have that. But you can draw a picture of *two* complex planes (side by side, say), take a simple figure in one plane (maybe a circle) and draw its *image* (that is, the squares of all the complex numbers on your circle) in the other plane. By looking at the images of simple figures like lines and circles, you get a feeling for how the squaring map works on the complex numbers. This is just the beginning. By *iterating* certain functions, you get intricate and beautiful pictures. This area of iterated complex maps is at the base of many computer algorithms for generating life-like graphics.

This chapter will contain a great deal of geometry for the "visual thinkers" among you.

Iterating a function means doing it over and over.

$$x \mapsto f(x) \mapsto f(f(x)) \mapsto f(f(f(x))) \mapsto \dots$$

Chapter 4. Combinations and Locks

There's one aspect of combinatorics that has been extremely useful in my teaching: the notion of a *combinatorial proof.* Often, students notice a pattern in their work that can be expressed as an identity. For example, students often come up with the fact that the sum of the entries in any row of Pascal's triangle is a power

Combinatorial proofs go by many names: "combinatorial arguments," "story proofs," "counting without counting," and "counting twice," among others. References [6, 19] are just two of the many books devoted to this approach.

Here, $\binom{n}{k}$ means the entry in the kth slot of the nth row of Pascal's triangle (where the numbering starts at 0).

of 2. This can be refined to an identity:

$$\binom{n}{0}+\binom{n}{1}+\binom{n}{2}+\cdots+\binom{n}{n}=2^n.$$

How do you prove such a thing? There are many ways, but a combinatorial proof amounts to finding a "story" in which you show that each side of the identity counts the same thing. In this case, the story involves picking subsets from a set of n things. How many ways are there to do this? Well, for each element there are two choices (in or out), so there are 2^n ways to do it. On the other hand, the chosen subset either has no elements, one element, two elements, ..., or n elements. And, there are $\binom{n}{2}$ ways to pick a 2-element subset, $\binom{n}{3}$ ways to pick a 3-element subset .... So, there are

If the details here are not perfectly clear to you, don't worry. We'll develop this example in detail later in the book.

$$\binom{n}{0}+\binom{n}{1}+\binom{n}{2}+\cdots+\binom{n}{n}$$

ways to pick a subset of any size. Since both sides of our identity count the same thing, they must be equal.

In this chapter, you'll practice using combinatorial proofs in algebra (the binomial theorem), geometry (counting paths), and, of course, combinatorics. This is a wonderful skill to develop.

We'll also look at a combinatorial problem that I've been using with high school students for over 15 years. The problem, determining the number of combinations on a certain kind of lock, always generates a great deal of interest, and it's a wonderful arena in which to apply one's skill at combinatorial proofs. And there's another bonus: some numbers that emerge in Chapter 1 when we are investigating difference tables make a surprise appearance here, in what seems like a completely different context.

Chapter 5. Sums of Powers

You may remember that there's a nice formula for the sum of the integers between 1 and n: $\frac{n(n+1)}{2}$. You may also know that there are formulas for the sums of other powers. Here's a list of the formulas for powers up to 8:

There's a legend involving the young Gauss that's often attached to this formula.

This means, for example, that

$$1^3+2^3+3^3+\cdots+n^3 = \frac{n^2(1+n)^2}{4}.$$

1. $\dfrac{n(1+n)}{2}$
2. $\dfrac{n(1+n)(1+2n)}{6}$
3. $\dfrac{n^2(1+n)^2}{4}$
4. $\dfrac{n(1+n)(1+2n)(-1+3n+3n^2)}{30}$
5. $\dfrac{n^2(1+n)^2(-1+2n+2n^2)}{12}$
6. $\dfrac{n(1+n)(1+2n)(1-3n+6n^3+3n^4)}{42}$
7. $\dfrac{n^2(1+n)^2(2-4n-n^2+6n^3+3n^4)}{24}$

Aren't these beautiful formulas? Look at the way common factors show up. We'll see in Chapter 5 how to look at the expressions in a way that makes them even more beautiful.

8. $\dfrac{n(1+n)(1+2n)(-3+9n-n^2-15n^3+5n^4+15n^5+5n^6)}{90}$

This chapter will investigate formulas like this—where they come from, how to generate them, and how to use them. For example, there's a general recursive method for finding any one of these formulas. And there's a method for generating them that comes right from the results of Chapter 1. In return, the formulas can be used to revisit the "finite differences" of Chapters 1 and 2, providing a particularly simple method for finding a polynomial that agrees with a table from its table of differences.

Some of the methods go back to Fermat and Pascal and make use of the results of Chapters 1 and 4. Others were developed by the Swiss mathematician Jakob Bernoulli.

Prerequisites

I'd like to say that there are none, except for an interest in doing mathematics, and that's almost true. But there are some things I'll assume, at least for some of the chapters:

- You should have a pretty solid background in high school mathematics, including an acquaintance with Pascal's triangle, the binomial theorem, proof by mathematical induction, and simple (arithmetic and geometric) series.
- In Chapter 2, it would be good if you know (or have once known) how to find the derivative of a polynomial and how to interpret the derivative as a formula for the slopes of tangents to its graph.
- In Chapter 3, you should know a little about complex numbers, the definitions of sine and cosine, and radian measure.
- In Chapter 4, we'll occasionally need to multiply matrices, and, although we assume no specific prerequisites, some of the ideas will be connected to basic ideas in linear algebra.
- Chapter 5 requires the ideas from Chapters 1, 2, and 4, and little else.

The last three chapters are somewhat independent, and each depends on the first two

Actually, some parts of Chapter 5 will be more meaningful if you've read Chapter 4.

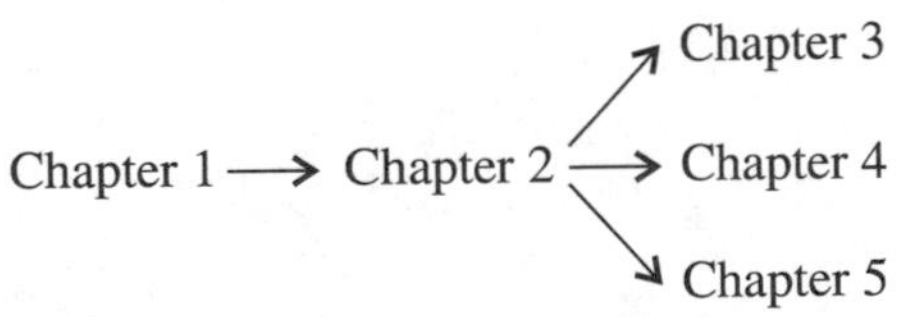

Chapter 1 ⟶ Chapter 2 ⟶ Chapter 3, Chapter 4, Chapter 5

But please, don't be scared off by all this. My advice is to dive into the chapters first and see if you can reconstruct the prerequisites when the need arises. If you need to consult a reference for some missing idea, you can find most of what you need in [15] and all of what you need in the additional references given at the end of each chapter.

1

Difference Tables and Polynomial Fits

Introduction

Many curricula ask students to look for patterns in data. Often, students are given a table like this:

Input	Output
0	3
1	8
2	13
3	18
4	23
5	28
6	33

And they are asked to find a "formula" that produces the table.

Let's hope they are asked to find *a* formula rather than *the* formula. There are, of course, many functions that agree with this table. Later, we'll see how to classify them (at least the ones given by polynomials).

Many beginning students don't look for a "closed form" solution at first. They don't ask "what can I do to 5 to get 28 so that when I do the same things to 6, I get 33?" Instead, they often subtract successive outputs and notice that the outputs "go up by 5 each time." It turns out that, if the differences go up by the same amount, there's a linear function (in this case, $f(x) = 5x + 3$) that fits the table. We'll see why in the next section.

Sometimes, the differences between successive outputs are *not* constant, but the successive differences of the differences *are* constant. When the "second differences" are constant, a *quadratic* function will fit the table. It's this phenomenon we want to investigate in this chapter. More precisely:

- We'll see why a constant mth difference means that a table can be matched by a polynomial of degree m.
- Conversely, we'll prove that a polynomial of degree m has constant mth difference.
- We'll come up with a pleasant and efficient method for *finding* a polynomial that matches a table from its difference table.

Actually, we'll look at fitting a polynomial to a table from three points of view:

1. In section 1.1, we'll introduce a method to which we'll return in Chapter 5. It is based on knowing formulas for the sums of powers.
2. In sections 1.2–1.3, we'll develop a method that goes back to Newton; it's based on a careful analysis of "difference tables" (defined below).
3. In section 1.7, we'll develop a technique attributed to Lagrange that generalizes an important theorem in arithmetic (the "Chinese remainder theorem") to polynomials.

So, by the end of the chapter, you'll have a nice collection of methods for fitting functions to data. And you'll have more:

The "important piece" is that the output for a particular input is the starting output plus the sum of all the differences up to that point. If this isn't clear, it will be shortly.

1. The ideas in section 1.1 form one of the important pieces underlying the fundamental theorem of calculus.
2. And, when we return to the method of section 1.1 in Chapter 5, we'll develop a collection of formulas that puts the famous

$$1+2+\cdots+n=\frac{n(n+1)}{2}$$

into a broader perspective.
3. The main method of this chapter (sections 1.2–1.3) will introduce you to the ideas of "discrete calculus." Discrete calculus is the analogue of ordinary calculus where *derivative* is replaced by *difference*. Discrete calculus is useful in all kinds of situations from solving difference equations (the analogue of differential equations) to finding closed forms for recursively defined functions. From a teaching perspective, it shows that many of the habits of mind underlying calculus are independent from notions of limit.

In the language of linear algebra, we'll choose different *bases* for our polynomials.

4. In sections 1.3–1.5, we'll see how writing polynomials in different forms can make certain kinds of information transparent. One of our methods will make it relatively simple to write down a polynomial that agrees with a table if, instead of writing it as a sum of monomials, we write it as a sum of other, equally simple, expressions.
5. And, in converting from one form to another, we'll meet some numbers that, like the entries in Pascal's triangle, show up all over mathematics.

After all, differences tell you something about how the function is changing.

6. This idea of subtracting successive numbers in a table, which seems to be ubiquitous among middle and high school students, turns out to be the germ of a general-purpose strategy for analyzing functions.

Let's get concrete about all this. First, a couple of introductory problems.

Problems

1. In a table like the one below, the Δ column is obtained by *subtracting* successive outputs. Confirm that each output can be obtained from the previous output by *adding* it to the adjacent entry in the Δ column.

We'll use this notation throughout the book: Δ (pronounced "delta") stands for "difference." Beside each output, we put in the Δ column the next output minus the current output. Beside 3 (the output for 0) goes $8 - 3 = 5$. Beside 8 (the output for 1) goes $13 - 8 = 5$. Beside 13 (the output for 2) goes $18 - 13 = 5$. All the differences are 5 in this example.

Input	Output	Δ
0	3	5
1	8	5
2	13	5
3	18	5
4	23	5
5	28	5
6	33	

Let's call this the "up and over" property of the table.

2. Show that, in an input-output table like the one below, every output is the sum of the output for 0 and the elements in the Δ column up to the number above the output.

We could say "The outputs are the running totals of the differences."

Input	Output	Δ
0	1	−2
1	−1	12
2	11	38
3	49	76
4	125	126
5	251	188
6	439	262
7	701	

Input	Output	Δ
0	1	−2
1	−1	12
2	11	38
3	49	76
4	125	126
5	251	188
6	439	262
7	701	

Input	Output	Δ
0	1	−2
1	−1	12
2	11	38
3	49	76
4	125	126
5	251	188
6	439	262
7	701	

Let's call this the "hockey stick" property of the table. So, every difference table has both the "up and over" and "hockey stick" properties.

1.1 Doing It with Sums

In this section, we'll look at one of the many methods for finding a polynomial function that agrees with a table. Let's return to the table on page 1.

Input	Output
0	3
1	8
2	13
3	18
4	23
5	28
6	33

The "first differences" are constant:

Input	Output	Δ
0	3	5
1	8	5
2	13	5
3	18	5
4	23	5
5	28	5
6	33	

For "constant difference" examples, you could reason like this:

An application of the hockey stick property (problem 2 on page 3).

I can get the 8 by taking $3+5$. And 13 is $8+5=(3+5)+5$. Ah, I just keep stacking up 5s:

Order of operations: $23+2\cdot 5$ means $23+(2\cdot 5)$.

$$\begin{aligned}
33 &= 28+5\\
&= (23+5)+5 = 23+2\cdot 5\\
&= (18+5)+2\cdot 5 = 18+3\cdot 5\\
&= (13+5)+3\cdot 5 = 13+4\cdot 5\\
&= (8+5)+4\cdot 5 = 8+5\cdot 5\\
&= (3+5)+5\cdot 5 = 3+6\cdot 5.
\end{aligned}$$

In general, to find the output at n, I start with 3 and add "n" 5s. So, the formula I want is $5n+3$. A function that agrees with the table is given by $f(n)=5n+3$.

This simple example already exhibits many features of the more general picture:

This is the "up and over" feature of difference tables (see problem 1 on page 3).

1. In a table like this, each output can be obtained from the previous output a by adding a and the entry in the Δ column to the right of a.
2. A constant Δ column means that a degree 1 (that is, a *linear*) polynomial will fit the table.
3. Finding a formula that agrees with a table is connected to adding up the numbers in the Δ column. (You can also call the Δ column "the first differences.")

Feature 1 is essential to everything we do with difference tables. It's true almost by definition (think of how you *get* the Δ column). You'll handle the details of establishing Feature 2 in the problem set. Feature 3 will come up over and over in this book; we'll deal with it in detail in Chapter 5.

Input	Output	Δ
0	3	5
1	8	5
2	13	5
3	18	5
4	23	5
5	28	5
6	33	

As we'll see later, there are many functions that agree with this table. In some sense, we're looking for the "simplest" function g that tabulates like this.

Let's look at another example. We have a mystery function g; what could a formula for g be?

n	$g(n)$
0	1
1	−2
2	1
3	10
4	25
5	46
6	73

Here's the table with *two* difference columns:

n	$g(n)$	Δ	Δ^2
0	1	−3	6
1	−2	3	6
2	1	9	6
3	10	15	6
4	25	21	6
5	46	27	
6	73		

We're using Δ^2 as shorthand for ΔΔ, the "second differences" or the "differences of the differences." This use of exponent notation is not very standard in precollege mathematics, but it's pretty common in post-secondary courses. One has to rely on context to know the operation to which the exponent refers; in this case, it's not repeated multiplication but repeated *application* of the Δ process.

So, Δ isn't constant. Ah, but *its* first differences *are* constant. Cover up the $g(n)$ column and make believe you were trying to find a formula for the Δ column; its outputs go up by the same amount each time, so there's a linear polynomial that produces the Δ column. In fact, it's $6n - 3$:

n	Δ	$\Delta^2 = \Delta\Delta$
0	−3	6
1	3	6
2	9	6
3	15	6
4	21	6
5	27	

One strategy in this game is to keep taking differences until you either run out of outputs or get a constant difference. Notice that the next difference column (the Δ^3 column) will be constant, too; it will be all 0s.

To get the entry in the Δ column for an input of n, start with −3 and add n 6s.

We might be able to make some progress on g now that we are equipped with this formula for the first differences.

n	$g(n)$	$\Delta = 6n - 3$
0	1	$-3 = 6 \cdot 0 - 3$
1	-2	$3 = 6 \cdot 1 - 3$
2	1	$9 = 6 \cdot 2 - 3$
3	10	$15 = 6 \cdot 3 - 3$
4	25	$21 = 6 \cdot 4 - 3$
5	46	$27 = 6 \cdot 5 - 3$
6	73	

Let's see where $g(5)$ comes from, using the hockey stick property, replacing each number in the Δ column by the calculation that led to it:

n	$g(n)$	$\Delta = 6n - 3$
0	1	-3
1	-2	3
2	1	9
3	10	15
4	25	21
5	46	27
6	73	

$$
\begin{aligned}
g(5) &= 1 + \boxed{-3} + \boxed{3} + \boxed{9} + \boxed{15} + \boxed{21} \\
&= 1 + \boxed{6 \cdot 0 - 3} + \boxed{6 \cdot 1 - 3} + \boxed{6 \cdot 2 - 3} + \boxed{6 \cdot 3 - 3} + \boxed{6 \cdot 4 - 3}
\end{aligned}
$$

Now group "like terms."

$$
\begin{aligned}
g(5) &= 1 + 6(0 + 1 + 2 + 3 + 4) + (-3 + -3 + -3 + -3 + -3) \\
&= 1 + 6(0 + 1 + 2 + 3 + 4) + 5(-3).
\end{aligned}
$$

Similarly (try it):

$$
\begin{aligned}
g(6) &= 1 + 6(0 + 1 + 2 + 3 + 4 + 5) + (-3 + -3 + -3 + -3 + -3 + -3) \\
&= 1 + 6(0 + 1 + 2 + 3 + 4 + 5) + 6(-3).
\end{aligned}
$$

Well, at least if $n = 0, \dots, 6$.

And, in general, a function g that agrees with the table will have the property that

$$
g(n) = 1 + 6(0 + 1 + 2 + 3 + 4 + 5 + \cdots + (n - 1)) + n(-3).
$$

Now, there's a famous formula for the sum of the first n integers:

$$0+1+2+3+\cdots+(n-1)=\frac{n(n-1)}{2}.$$

If you believe that, we have

You can check that this agrees with the table.

$$\begin{aligned} g(n) &= 1+6\big(0+1+2+3+4+5+\cdots+(n-1)\big)+n\cdot -3 \\ &= 1+6\left(\frac{n(n-1)}{2}\right)-3n \\ &= 3n^2-6n+1. \end{aligned}$$

If you work through some more examples, it becomes clear that a table with a constant *second* difference can be matched by a polynomial of degree 2.

But proceeding along the lines we've started here will take some work. A constant first difference required us to add up n identical constants. A constant second difference required us to add up the first n integers. It turns out that a constant third difference requires us to add up the first n *squares*. And in general, a constant difference in the mth column requires that you are able to add up the first n $(m-1)$st powers. We'll get formulas for such sums in Chapter 5, and we'll return to this approach then, finishing the story. This will give us a perfectly serviceable method for "resolving" difference tables, once we have nice formulas for sums of powers.

The sum of constants ($m=0$) is a linear function, the sum of a linear function ($m=1$) is a quadratic, the sum of squares turns out to be a cubic, and so on. If we knew this for sure, we'd be able to prove that, if the mth differences are constant, there's a polynomial of degree m that fits the table.

Ways to think about it

We started down a natural path of investigation and came up with a promising method for finding a polynomial that agrees with a table: Take the differences until you get a constant, then successively find a formula for each column (by summing the column to the right of it). Trouble is, these formulas get harder to find as the power goes up.

This happens all the time in mathematics research. The solution of one problem is reduced to the solution of another. Sometimes, the second problem is easier to solve, sometimes it's not. When it's not, as in our case, there are two things one can do:

- Try another approach.
- Dig in, pour some coffee, and work on the "reduced" problem.

We'll do both in this book. In the next section, we'll look at a slightly different approach for fitting polynomials to data that is simple, computationally feasible, and fun. But the question of summation formulas is too intriguing to leave, and it has applications beyond finding functions that agree with tables. So, in Chapter 5, we'll return to the topic of summations, and we'll get the formulas we need to finish off the approach we started here.

Have we solved the problem of finding a function that agrees with a table? Well, we've reduced it to doing something else (finding summation formulas). Sometimes, people in this position say they've solved the problem "in principle." If nothing else, that makes you feel better.

Problems

3. Find a polynomial function that agrees with each table.

(a)

Input	Output
0	1
1	4
2	7
3	10
4	13
5	16

(b)

Input	Output
0	3
1	6
2	9
3	12
4	15
5	18

(c)

Input	Output
0	3
1	4
2	8
3	15
4	25
5	38

(d)

Input	Output
0	5
1	6
2	10
3	17
4	27
5	40

(e)

Input	Output
0	5
1	7
2	12
3	20
4	31
5	45

(f)

Input	Output
0	10
1	14
2	24
3	40
4	62
5	90

4. Find a function that agrees with this table:

Input	Output	Δ
0	c	d
1		d
2		d
3		d
4		d
5		

Legend has it that the young Gauss figured out this formula and used it to add up the integers between 1 and 100 in a couple seconds. There's also a rumor that he used no such formula—he just added the numbers in his head.

If you get stuck, you can find proofs in [15] or on the web. You can always use mathematical induction on these things, but try to find a proof that shows how someone might have come upon the formulas.

5. (a) While you are at it, show that the famous formula for the sum of the first n integers:

$$0+1+2+3+\cdots+(n-1)=\frac{n(n-1)}{2}.$$

(b) While you are at it, show that

$$0^2+1^2+2^2+3^2+\cdots+(n-1)^2=\frac{n(n-1)(2n-1)}{6}.$$

6. Find a function that agrees with this table:

Input	Output	Δ	Δ^2
0	3	4	2
1			2
2			2
3			2
4			
5			

You might want to fill in the table first. Is there only one table that agrees with these "'boundary conditions?"

7. Find a function that agrees with this table:

Input	Output	Δ	Δ^2
0	a	b	c
1			c
2			c
3			c
4			
5			

Recast your solution into the statement of a theorem.

8. Find a possible formula for $f(n)$:

Maybe use the result from problem 5?

n	$f(n)$
0	1
1	−1
2	11
3	49
4	125
5	251
6	439
7	701

9. **(a)** Suppose you knew that the formula for the sum of the first n cubes is a polynomial in n of degree 4. That is:

$$0^3 + 1^3 + 2^3 + \cdots + n^3 = a_4n^4 + a_3n^3 + a_2n^2 + a_1n + a_0.$$

Use this to prove that, if the 4th differences in a table are constant, a polynomial of degree 4 will fit the table exactly.

(b) Suppose you knew that the formula for the sum of the first n $(m-1)$th powers is a polynomial in n of degree m. Use this to prove that, if the mth differences in a table are constant, a polynomial of degree m will fit the table exactly.

It really is true. In Chapter 5, we'll derive a whole sequence of formulas like this:

m	$0^m + 1^m + \cdots + (n-1)^m$
0	$n-1$
1	$\frac{n(n-1)}{2}$
2	$\frac{n(n-1)(2n-1)}{6}$
3	$\frac{n^2(n-1)^2}{4}$

1.2 Doing It with Differences

When looking for an algorithm that solves a class of problems, it's a good idea to work out some numerical examples, concentrating on the rhythm of the calculations rather than on the numbers with which you work. This next discussion is a good example of that.

Let's switch perspective a little. The goal is still to find a simple polynomial function that agrees with a table, but, instead of working from "right to left" in these difference tables, let's work from "bottom to top." An example will help show what we mean. Here's a table (the one from problem 8), together with all the difference columns, until we get a constant:

Input	Output	Δ	Δ^2	Δ^3
0	1	−2	14	12
1	−1	12	26	12
2	11	38	38	12
3	49	76	50	12
4	125	126	62	12
5	251	188	74	
6	439	262		
7	701			

n	$f(n)$	Δ	Δ^2	Δ^3
0	1	−2	14	12
1	−1	12	26	12
2	11	38	38	12
3	49	76	50	12
4	125	126	62	12
5	251	188	74	
6	439	262		
7	701			

The third difference is constant. Suppose we want to find a function f (a polynomial, say) that agrees with this table. Let's just see what we can get from the "up and over" feature of the table.

Notice that, by the way we built difference tables, *every* entry in the interior of the table is the sum of its "up and over." $49 = 11 + 38$, $62 = 50 + 12$, $262 = 188 + 74$, ... So, we can take any entry in the $f(n)$ column, replace it by its up and over, replace these two numbers by *their* up and overs, and keep moving up the table toward the first row. Let's try it with $f(3)$:

$$\begin{aligned} f(3) &= 49 \\ &= 11 + 38 \\ &= (-1 + 12) + (12 + 26) = -1 + 2 \cdot 12 + 26 \\ &= (1 + -2) + 2 \cdot (-2 + 14) + 14 + 12 = 1 + 3 \cdot (-2) + 3 \cdot 14 + 12. \end{aligned}$$

There are many ways to write $f(3)$ as a sum of numbers. Here, we're trying to follow the *form* of how you'd chase $f(3)$ up to the 0th row. Try it yourself, writing each step in terms of numbers in the table.

Look again, with some emphasis added and details suppressed:

$$\begin{aligned} f(3) &= \mathbf{1} \cdot 49 \\ &= \mathbf{1} \cdot 11 + \mathbf{1} \cdot 38 \\ &= \mathbf{1} \cdot -1 + \mathbf{2} \cdot 12 + \mathbf{1} \cdot 26 \\ &= \mathbf{1} \cdot 1 + \mathbf{3} \cdot (-2) + \mathbf{3} \cdot 14 + \mathbf{1} \cdot 12. \end{aligned}$$

Could it be? It looks as if Pascal's triangle is rearing its head once again. Let's try it for $f(4)$.

$$
\begin{aligned}
f(4) &= \mathbf{1}\cdot 125\\
&= \mathbf{1}\cdot 49 + \mathbf{1}\cdot 76\\
&= \mathbf{1}\cdot(11+38) + \mathbf{1}\cdot(38+38)\\
&= \mathbf{1}\cdot 11 + \mathbf{2}\cdot 38 + \mathbf{1}\cdot 38\\
&= \mathbf{1}\cdot(-1+12) + \mathbf{2}\cdot(12+26) + \mathbf{1}\cdot(26+12)\\
&= \mathbf{1}\cdot -1 + \mathbf{3}\cdot 12 + \mathbf{3}\cdot 26 + \mathbf{1}\cdot 12\\
&= \mathbf{1}\cdot(1+-2) + \mathbf{3}\cdot(-2+14) + \mathbf{3}\cdot(14+12) + \mathbf{1}\cdot(12+0)\\
&= \mathbf{1}\cdot 1 + \mathbf{4}\cdot(-2) + \mathbf{6}\cdot 14 + \mathbf{4}\cdot 12 + \mathbf{1}\cdot 0.
\end{aligned}
$$

In order to make this work, we had to add more columns to the difference table:

It *is* true that $\Delta^4 = 0$ for this table, right?

n	$f(n)$	Δ	Δ^2	Δ^3	Δ^4	Δ^5
0	1	−2	14	12	0	0
1	−1	12	26	12	0	0
2	11	38	38	12	0	0
3	49	76	50	12	0	
4	125	126	62	12		
5	251	188	74			
6	439	262				
7	701					

Facts and Notation

For our current discussion about Pascal's triangle, we need three things: some notation for its entries, an algorithm for generating the triangle, and an explicit formula for the entries in terms of factorials. To that end:

Pascal's triangle has many wonderful properties and is connected to a host of mathematical phenomena. For more details and a complete development, see, for example, [18].

There are several different ways to get the numbers in Pascal's triangle. It's by no means obvious that all the ways produce the same numbers, but they do. We'll list several ways here. The proofs that all these methods yield the same results can be carried out by induction or by other means (see [15], for example). If you have time, try proving some of the equivalences for yourself. We'll use $\binom{n}{k}$ to stand for the entry in the nth row and kth position in the triangle. Depending on your previous experiences, this notation might conjure up any one of the following ways to think about Pascal's triangle. Think of *that* way as a definition and think about the rest of the ways as theorems.

- *Pascal's triangle is a recursively generated number pattern.*

$$\binom{0}{0} = 1$$
$$\binom{1}{0} = 1 \quad \binom{1}{1} = 1$$
$$\binom{2}{0} = 1 \quad \binom{2}{1} = 2 \quad \binom{2}{2} = 1$$
$$\binom{3}{0} = 1 \quad \binom{3}{1} = 3 \quad \binom{3}{2} = 3 \quad \binom{3}{3} = 1$$
$$\binom{4}{0} = 1 \quad \binom{4}{1} = 4 \quad \binom{4}{2} = 6 \quad \binom{4}{3} = 4 \quad \binom{4}{4} = 1$$
$$\binom{5}{0} = 1 \quad \binom{5}{1} = 5 \quad \binom{5}{2} = 10 \quad \binom{5}{3} = 10 \quad \binom{5}{4} = 5 \quad \binom{5}{5} = 1$$

Notice that the numbering starts with 0. So, for example,

$$\binom{6}{0} = 1 \quad \text{and} \quad \binom{6}{1} = 6.$$

The rule for *generating* the triangle in this way can be stated in words:

This "two-term recurrence" (every number is the sum of the two above it) is satisfied by many mathematical phenomena.

Each row starts and ends with 1, and any interior element is the sum of the two above it

or in symbols:

$$\binom{n}{k} = \begin{cases} 1 & \text{if } k = 0 \text{ or if } k = n \text{ (each row starts and ends with 1), and} \\ \binom{n-1}{k-1} + \binom{n-1}{k} & \text{if } 0 < k < n \text{ (an interior element is the sum of the two above it).} \end{cases}$$

- *The entries in Pascal's triangle count subsets.* Suppose you have a set of five elements, say $\{A, B, C, D, E\}$. How many three-element subsets are there? There are $\binom{5}{3} = 10$. Here they are:

$$\{A, B, C\}, \{A, B, D\}, \{A, B, E\}, \{A, C, D\}, \{A, C, E\}, \{A, D, E\},$$
$$\{B, C, D\}, \{B, C, E\}, \{B, D, E\}, \{C, D, E\}.$$

This is called the binomial theorem, and it is proved in [15].

- *The entries in Pascal's triangle are the coefficients in $(a+b)^n$.* The entries in the nth row are the coefficients in the expansion of $(a+b)^n$. More precisely,

$$\begin{aligned}(a+b)^n &= \binom{n}{0}a^n + \binom{n}{1}a^{n-1}b + \binom{n}{2}a^{n-2}b^2 + \cdots + \binom{n}{n-2}a^2b^{n-2} \\ &\quad + \binom{n}{n-1}ab^{n-1} + \binom{n}{n}b^n \\ &= \sum_{k=0}^{n} \binom{n}{k} a^{n-k}b^k.\end{aligned}$$

- *The entries in Pascal's triangle are quotients of factorials.* There's an explicit formula for $\binom{n}{k}$ in terms of factorials:

$$\binom{n}{k} = \frac{n!}{k!(n-k)!}.$$

So, for example,

$$\binom{12}{5} = \frac{12!}{5!7!}$$
$$= \frac{12 \cdot 11 \cdot 10 \cdot 9 \cdot 8 \cdot 7 \cdot 6 \cdot 5 \cdot 4 \cdot 3 \cdot 2 \cdot 1}{5 \cdot 4 \cdot 3 \cdot 2 \cdot 1 \times 7 \cdot 6 \cdot 5 \cdot 4 \cdot 3 \cdot 2 \cdot 1}$$
$$= \frac{12 \cdot 11 \cdot 10 \cdot 9 \cdot 8}{5 \cdot 4 \cdot 3 \cdot 2 \cdot 1}$$
$$= 792.$$

Making this factorial formula hold for the cases $k = 0$ and $k = n$ is one of the reasons for defining 0! to be 1.

- *The entries in Pascal's triangle are rational expressions.* Sometimes, it's useful to do the cancellations in the factorial expression and write $\binom{n}{k}$ as a product of factors:

$$\binom{n}{k} = \frac{n(n-1)(n-2)(n-3)\cdots(n-k+1)\,(n-k)\cdots 1}{k!\,(n-k)!}$$
$$= \frac{n(n-1)(n-2)(n-3)\cdots(n-k+1)}{k!}.$$

One advantage of this expression over all the others is that n can be any number here. In a sense, this expression extends the formula for entries in Pascal's triangle from integers to real (or even complex) numbers. This will be important in section 1.3.

The symbol $\binom{\pi}{3}$ doesn't make sense in the context of any of the other methods. But here, it's just

$$\frac{\pi(\pi-1)(\pi-2)}{6}.$$

So, it looks as if we have a general method for calculating $f(n)$ for all the inputs in our table on page 11. Let's see if we can turn this into an explicit (polynomial) formula for f. First of all, we can ignore the Δ^e columns for $e > 3$, because they are all 0. So far we have:

$$f(3) = 49 \quad = \mathbf{1} \cdot 1 + \mathbf{3} \cdot (-2) + \ \ \mathbf{3} \cdot 14 + \ \ \mathbf{1} \cdot 12,$$
$$f(4) = 125 = \mathbf{1} \cdot 1 + \mathbf{4} \cdot (-2) + \ \ \mathbf{6} \cdot 14 + \ \ \mathbf{4} \cdot 12,$$
$$f(5) = 251 = \mathbf{1} \cdot 1 + \mathbf{5} \cdot (-2) + \mathbf{10} \cdot 14 + \mathbf{10} \cdot 12,$$
$$f(6) = 439 = \mathbf{1} \cdot 1 + \mathbf{6} \cdot (-2) + \mathbf{15} \cdot 14 + \mathbf{20} \cdot 12,$$
$$f(7) = 701 = \mathbf{1} \cdot 1 + \mathbf{7} \cdot (-2) + \mathbf{21} \cdot 14 + \mathbf{35} \cdot 12.$$

Many discoveries in algebra are made through a careful and mechanical analysis of a calculation, concentrating on the *form* of the calculation rather than on its *value*. Notice that the equations start with $f(3)$ and not $f(0)$. Why?

And, if we determined further outputs for f by "pulling out" from the constant difference of 12, the same method would work for $f(8)$, $f(9)$, and $f(10)$: If you snake back through the difference table, winding your way to the top row by replacing each number by its "up and over," you'll get a combination of the elements in the first row, and the coefficients will be precisely the entries in the "input" row of Pascal's triangle. So, it looks as if we could calculate $f(n)$ by the rule

$$f(n) = \binom{n}{0} \cdot 1 + \binom{n}{1} \cdot (-2) + \binom{n}{2} \cdot 14 + \binom{n}{3} \cdot 12.$$

By "pulling out from the constant difference of 12," we mean adding a few more 12s in the Δ^3 column and then filling the table in, right to left, using the "up and over" rule.

This certainly works if n is 3, 4, 5, 6, or 7. Before we worry about the details, let's see if this will help us get a general formula for f. Since we're looking for a polynomial, let's get rid of the $\binom{n}{k}$ expressions and replace them with their algebraic equivalents:

$$f(n) = 1 \cdot 1 + n \cdot (-2) + \frac{n(n-1)}{2} \cdot 14 + \frac{n(n-1)(n-2)}{6} \cdot 12.$$

Oh, look at this. A little algebra (by hand or with a CAS) can turn this into a nice cubic:

$$f(n) = 2n^3 + n^2 - 5n + 1.$$

You can check that this function agrees with the table.

Ways to think about it

There are many loose ends here. For one, the formula that gave us the nice cubic:

$$f(n) = \binom{n}{0} \cdot 1 + \binom{n}{1} \cdot (-2) + \binom{n}{2} \cdot 14 + \binom{n}{3} \cdot 12 \qquad (*)$$

doesn't work if n is 0, 1, or 2, because, for example, we have no meaning for $\binom{2}{3}$. In fact, chasing the "up and over" property for 0, 1, and 2 up to the top row produces what look like different equations, "truncated" versions of $(*)$:

$$f(0) = \binom{0}{0} \cdot 1,$$

$$f(1) = \binom{1}{0} \cdot 1 + \binom{1}{1} \cdot (-2),$$

$$f(2) = \binom{2}{0} \cdot 1 + \binom{2}{1} \cdot (-2) + \binom{2}{2} \cdot 14.$$

So, from now on, $\binom{3}{5} = 0$, $\binom{5}{21} = 0$, $\binom{6}{7} = 0$, and $\binom{0}{1} = 0$. This even makes sense using the "subset" interpretation of $\binom{2}{3}$: How many 3-element subsets does a 2-element set have? None. In some investigations, it's even useful to extend Pascal's triangle to negative numbered rows.

But formula $(*)$ says that these equations should be

$$f(0) = \binom{0}{0} \cdot 1 + \binom{0}{1} \cdot (-2) + \binom{0}{2} \cdot 14 + \binom{0}{3} \cdot 12,$$

$$f(1) = \binom{1}{0} \cdot 1 + \binom{1}{1} \cdot (-2) + \binom{1}{2} \cdot 14 + \binom{1}{3} \cdot 12,$$

$$f(1) = \binom{2}{0} \cdot 1 + \binom{2}{1} \cdot (-2) + \binom{2}{2} \cdot 14 + \binom{2}{3} \cdot 12.$$

Ah, but if we look at $\binom{n}{3}$ as a rational expression in n (as in the last bullet on page 13), $\binom{2}{3}$, for example, and more generally $\binom{n}{k}$ with $k > n$, make sense and are, in fact, 0:

$$\binom{2}{3} = \frac{2(2-1)(2-2)}{3!} = 0.$$

Let's do that from now on: Whenever we see $\binom{n}{k} = 0$ with $k > n$, we'll think of the rational expression form, and replace it by 0.

Problems

10. Recall this table from page 11:

n	$f(n)$	Δ	Δ^2	Δ^3	Δ^4	Δ^5
0	1	−2	14	12	0	0
1	−1	12	26	12	0	0
2	11	38	38	12	0	0
3	49	76	50	12	0	
4	125	126	62	12		
5	251	188	74			
6	439	262				
7	701					

(a) Show that

$$\begin{aligned} f(5) &= \mathbf{1}\cdot 1 + \mathbf{5}\cdot(-2) + \mathbf{10}\cdot 14 + \mathbf{10}\cdot 12 + \mathbf{5}\cdot 0 + \mathbf{1}\cdot 0 \\ &= \mathbf{1}\cdot 1 + \mathbf{5}\cdot(-2) + \mathbf{10}\cdot 14 + \mathbf{10}\cdot 12. \end{aligned}$$

(b) Show similar results for $f(6)$ and $f(7)$ too.

It might be handy to pull out a copy of Pascal's triangle for reference.

11. Use the method of this section to find a polynomial that agrees with each table.

(a)

Input	Output
0	1
1	5
2	25
3	79
4	185
5	361
6	625
7	995

(b)

Input	Output
0	−1
1	0
2	−5
3	−22
4	−57
5	−116
6	−205
7	−330

(c)

Input	Output
0	1
1	0
2	7
3	26
4	63
5	124
6	215
7	342

(d)

Input	Output
0	−1
1	4
2	43
3	194
4	583
5	1384
6	2819
7	5158

(e)

Input	Output
0	2
1	9
2	26
3	53
4	90
5	137
6	194
7	261

(f)

Input	Output
0	1
1	6
2	63
3	364
4	1365
5	3906
6	9331
7	19608

12. If a table has 12 inputs, for $n = 0, \ldots, 11$, how many Δ columns can be created?

For fun, fill in the rest of the table and check your answer.

13. Find a polynomial function that agrees with this table:

Input	Output	Δ	Δ^2	Δ^3
0	3	−4	5	2
1				2
2				
3				
4				
5				

Is there more than one possible answer to this? Is there more than one polynomial function of a given degree that agrees with the table?

14. Find a polynomial function that agrees with this table:

Input	Output	Δ	Δ^2	Δ^3
0	a	b	c	d
1				d
2				
3				
4				
5				

Is there more than one possible answer to this?

15. Prove Theorem 1 below:

Theorem 1. *Suppose we have a table with inputs* $0, \ldots, m$*:*

Input	*Output*	Δ	Δ^2	Δ^3	$\ldots$	Δ^m
0	a_0	a_1	a_2	a_3	$\ldots$	a_m
1						
2						
3						
4						
5						
6						
$\vdots$						
m						

If f is a function that agrees with the table, then for $0 \le n \le m$:

$$f(n) = \sum_{k=0}^{m} a_k \binom{n}{k}.$$

1.3 Finding a Formula: Combinatorial Polynomials

Theorem 1 is the basis for a simple technique, called *Newton's Difference Formula*, to find a polynomial—of smallest degree, even—that agrees with any finite input-output table (whose inputs are consecutive integers starting with 0). To do this, we introduce a new collection of polynomials that "extend" the binomial coefficients.

For example, is there a polynomial function g that has the property that when you take $g(n)$ (n an integer), you get $\binom{n}{3}$? The answer is deceptively simple. To think about $\binom{n}{3}$, look at the rational expression:

$$\binom{n}{3} = \frac{n(n-1)(n-2)}{6}.$$

So, let

$$g(x) = \frac{x(x-1)(x-2)}{6}.$$

Then g is certainly a polynomial function, so you can take g of any real (or even complex) number. In fact, expanding the fraction, we get $g(x) = \frac{1}{6}x^3 - \frac{1}{2}x^2 + \frac{1}{3}x$ so g is given by a cubic polynomial with rational coefficients. And looking at g in its factored form, one sees that $g(n) = \binom{n}{3}$ for any integer n. So, we have "extended" the binomial coefficient $\binom{n}{3}$ by a cubic polynomial.

More generally, we can define a sequence of polynomials that extends all the binomial coefficients.

We're using 3 here just for the sake of example. Any other positive integer would do.

In[15], we often use another device to define functions, especially when we don't care about the function's name. Instead of writing "$f(x) = x^2 + 1$", we often write "$x \mapsto x^2 + 1$" and say, "x maps to $x^2 + 1$."

Extension in this sense means "extend the function to values not in its domain." Until now, $n \mapsto \binom{n}{3}$ had the nonnegative integers as a domain (remember that $\binom{n}{3} = 0$ if $n \in \{0, 1, 2\}$). The function g agrees with $n \mapsto \binom{n}{3}$ for nonnegative integers but is defined on all of $\mathbb{R}$.

Definition Suppose k is a nonnegative integer. The kth *combinatorial polynomial*, $\binom{x}{k}$, is defined by the rule

$$\binom{x}{k} = \frac{x(x-1)(x-2)(x-3)\cdots(x-k+1)}{k!}$$

By convention, we take $\binom{x}{0}$ to be 1.

Ways to think about it

There's a suggestive piece of notation here. We use "$\binom{x}{k}$" to stand for the kth combinatorial polynomial. It extends the binomial coefficients $\binom{n}{k}$ to $\mathbb{R}$. You can think "when I replace x by n in $\binom{x}{k}$, I get $\binom{n}{k}$." But that's a *theorem*, made transparent by the notation.

These were generated with a CAS.

Here are a few of the combinatorial polynomials:

k	$\binom{x}{k}$
0	1
1	x
2	$\frac{-x+x^2}{2}$
3	$\frac{2x-3x^2+x^3}{6}$
4	$\frac{-6x+11x^2-6x^3+x^4}{24}$
5	$\frac{24x-50x^2+35x^3-10x^4+x^5}{120}$
6	$\frac{-120x+274x^2-225x^3+85x^4-15x^5+x^6}{720}$
7	$\frac{720x-1764x^2+1624x^3-735x^4+175x^5-21x^6+x^7}{5040}$

Before we investigate the properties of the various $\binom{x}{k}$, let's state one of the main results of this chapter:

Theorem 2. (Newton's Difference Formula.) *Suppose we have a table whose inputs are the integers between 0 and m:*

Input	*Output*	Δ	Δ^2	Δ^3	$\dots$	Δ^m
0	a_0	a_1	a_2	a_3	$\dots$	a_m
1						
2						
3						
4						
5						
6						
$\vdots$						
m						

A polynomial function that agrees with the table is

$$f(x) = \sum_{k=0}^{m} a_k \binom{x}{k}.$$

Furthermore, f has degree at most m.

Proof The fact that f has degree at most m comes from the explicit formula for $\binom{x}{k}$:

$$\binom{x}{k} = \frac{x(x-1)(x-2)(x-3)\cdots(x-k+1)}{k!}.$$

There are k factors on top, so this has degree k. Since f is a sum of constant multiples of the $\binom{x}{k}$, its degree is no greater than the largest degree of any summand, that is, of $\binom{x}{m}$.

To see that the f given in the theorem does what it's supposed to do (that is, agrees with the table), use Theorem 1 and the fact that, if n is substituted for x in $\binom{x}{k}$, the result is $\binom{n}{k}$. ■

So, now fitting a polynomial to a table becomes a completely mechanical process.

Example: Here's the table from problem 11f on page 15.

The polynomial resulting from Newton's difference formula is quite simple in this case. That's because we wanted a simple example. Usually, the answers aren't so nice. But the method always works.

Input	Output
0	1
1	6
2	63
3	364
4	1365
5	3906
6	9331
7	19608

Let's complete its difference table:

We could *really* complete the difference table, adding the columns of 0s, but that wouldn't change the resulting function, right?

Input	Output	Δ	Δ^2	Δ^3	Δ^4	Δ^5
0	1	5	52	192	264	120
1	6	57	244	456	384	120
2	63	301	700	840	504	120
3	364	1001	1540	1344	624	
4	1365	2541	2884	1968		
5	3906	5425	4852			
6	9331	10277				
7	19608					

So, with no further fuss, here's a (degree 5) polynomial function that agrees with the table:

$$
\begin{aligned}
f(x) &= 1 \cdot \binom{x}{0} + 5 \cdot \binom{x}{1} + 52 \cdot \binom{x}{2} + 192 \cdot \binom{x}{3} + 264 \cdot \binom{x}{4} + 120 \cdot \binom{x}{5} \\
&= 1 + 5x + 52\frac{x(x-1)}{2} + 192\frac{x(x-1)(x-2)}{6} \\
&\quad + 264\frac{x(x-1)(x-2)(x-3)}{24} \\
&\quad + 120\frac{x(x-1)(x-2)(x-3)(x-4)}{120} \\
&= 1 + x + x^2 + x^3 + x^4 + x^5
\end{aligned}
$$

The last simplification was done with a little help from a CAS.

Notice that $f(x)$ can also be written as

$$\frac{x^6 - 1}{x - 1}.$$

Problems

These tables are from problem 11 on page 15. A CAS will help greatly here.

16. Use the result of Theorem 2 to find a polynomial that agrees with each table.

(a)

Input	Output
0	1
1	5
2	25
3	10
4	79
5	185
6	361
7	995

(b)

Input	Output
0	−1
1	0
2	−5
3	−22
4	−57
5	−116
6	−205
7	−330

(c)

Input	Output
0	−1
1	0
2	7
3	26
4	63
5	124
6	215
7	342

(d)

Input	Output
0	−1
1	4
2	43
3	194
4	583
5	1384
6	2819
7	5158

(e)

Input	Output
0	2
1	9
2	26
3	53
4	90
5	137
6	194
7	261

(f)

Input	Output
0	1
1	6
2	63
3	364
4	1365
5	3906
6	9331
7	19608

17. Fran hands you a table and asks you to find a function that agrees with it.

Input	Output
0	−5
1	−6
2	7
3	166
4	843
5	2770

Find a function f that will agree with Fran's table. To save you some work, here's the completed difference table:

Input	Output	Δ	Δ^2	Δ^3	Δ^4	Δ^5
0	−5	−1	14	132	240	120
1	−6	13	146	372	360	
2	7	1591	518	732		
3	166	677	1250			
4	843	1927				
5	2770					

18. Fran forgot one input in the table in problem 17; the real table is:

Input	Output
0	−5
1	−6
2	7
3	166
4	843
5	2770
6	4999

Does your function f from problem 17 still work?

- Find a function g that agrees with this new table.
- Let $h(x) = f(x) - g(x)$. Tabulate h between 0 and 7.

"Tabulate h" means "make a table for h."

19. Graph the functions $x \mapsto \binom{x}{k}$ for $k = \{0, \ldots, 10\}$ between $x = -1$ and $x = 10$.

When you are done, put them on the same axes. Pretty, no?

Problems 20 and 21 look at the uniqueness of the function produced by Newton's difference formula.

Use anything you need from high school algebra to make the argument.

20. Algebra Review. Suppose that f and g are two polynomial functions of degree at most m. Show that if f and g agree at $m+1$ inputs, they are identical polynomials.

21. Show that the function you get using Theorem 2 is the unique polynomial function of smallest degree that agrees with a given table.

1.4 Making It Formal: The Δ Operator

In this section, we'll refine our notion of Δ in a way that will give us a useful tool for the rest of this book. We'll also shift perspective a bit. Up until now, we had a table and we wanted to fit it with a function. Now we imagine starting with the function and we see what happens if we "Δ it."

So far, we've been using Δ to stand for an operation we perform on tables (on columns of numbers, really). There's another way to think about it. You can think of Δ as an operation on *functions*. For example, suppose $f(x) = x^2 + 1$. The table and its first differences look like this:

n	$f(n)$	Δ
0	1	1
1	2	3
2	5	5
3	10	7
4	17	9
5	26	11
6	37	

The Δ column defines a function in its own right. In fact, it's not hard to see that it can be taken to be $x \mapsto 2x+1$. So, you can think of Δ as a process that transforms one function ($x \mapsto x^2+1$) to another one ($x \mapsto 2x+1$). We could write it this way:

$$\Delta(x \mapsto x^2 + 1) = x \mapsto 2x + 1.$$

Example: Make a table for the function $x \mapsto x^3 + 3x$ and look at its first differences.

x	$x \mapsto x^3 + 3x$	Δ
0	0	4
1	4	10
2	14	22
3	36	40
4	76	64
5	140	94
6	234	

So, what's a formula for $\Delta(x \mapsto x^3+3x)$? You can poke around to find one (or, use the method of the previous section). But you could also reason like this:

This is the same thing as asking for a formula that fits the Δ column in the table.

First of all, name the thing you are "differencing": Let $f(x) = x^3+3x$. Then you want a formula for $\Delta(f)$. Use the algebraic habit of "form not value" and look at how you'd get various values of $\Delta(f)$.

It takes some getting used to to see "$\Delta(f)$" and think "a function." If it makes it easier, replace $\Delta(f)$ by some letter like g for awhile.

$$\Delta(f)(0) = f(1) - f(0)$$

$$\Delta(f)(1) = f(2) - f(1)$$

$$\Delta(f)(2) = f(3) - f(2)$$

$$\Delta(f)(3) = f(4) - f(3)$$

$$\Delta(f)(4) = f(5) - f(4)$$

Ah. There we go. We can do it once and for all and be done with it:

When we look at the table, we think of x as an integer. But x could be anything in the domain of f.

$$\begin{aligned}\Delta(f)(x) &= f(x+1) - f(x)\\ &= \left((x+1)^3 + 3(x+1)\right) - (x^3+3x)\\ &= 3x^2 + 3x + 4.\end{aligned}$$

So,

$$\Delta(x \mapsto x^3+3x) = x \mapsto 3x^2+3x+4.$$

If you check out this formula against the Δ column, you'll see an exact match.

So, if we think of Δ as an operator that takes one function and produces another, we can make a definition:

Definition If f is any function defined on, say, the real numbers, $\Delta(f)$ is the new function defined by

We realize that this is getting pretty notation-heavy. This is it for a while. Almost.

$$\Delta(f)(x) = f(x+1) - f(x)$$

Ways to think about it

The operator Δ is an example of what computer scientists call a "higher-order" function: a function whose inputs and outputs are themselves functions. You may have met other higher-order functions in calculus: The *derivative* can be looked at as a machine that turns functions into functions. In linear algebra, taking the *transpose* can be thought of as a way to change linear transformations into other linear transformations.

What's confusing sometimes is when we write "$\Delta(f)(8)$." To parse this, think, "I apply Δ to f and get a function. I apply this new function to 8 (and get a number, probably)." We could add more parentheses to make

(*continued*)

Alternatively, if the world you care about is polynomials, you can think of Δ as an operation that assigns a polynomial to another polynomial, where you think of a polynomial as a formal expression.

the order explicit, writing something like

$$(\Delta(f))(8),$$

but this only adds notational clutter with no guarantee that it makes the ideas clearer.

Here's the last little bit of notation: $\Delta^2(f)(n)$ means $\Delta(\Delta(f))(n)$ etc. Let's let $\Delta^0(f)$ be f itself.

So, now we can write our difference tables in a way that makes all the column headings consistent: For the function $f(x) = 2x^3 + x^2 - 5x + 1$, we have (same table, new look):

n	$f(n)$	$\Delta(f)(n)$	$\Delta^2(f)(n)$	$\Delta^3(f)(n)$	$\Delta^4(f)(n)$	$\Delta^5(f)(n)$
0	1	−2	14	12	0	0
1	−1	12	26	12	0	0
2	11	38	38	12	0	0
3	49	76	50	12	0	
4	125	126	62	12		
5	251	188	74			
6	439	262				
7	701					

We'll see even more applications of this way of thinking about Δ as a higher-order function in the next section.

This section would be a huge waste of time if all it did was make column headings consistent. But it does more: It allows us to express any *row* of the difference table generically, for any function f:

n	$f(n)$	$\Delta(f)(n)$	$\Delta^2(f)(n)$	$\Delta^3(f)(n)$	$\Delta^4(f)(n)$	$\Delta^5(f)(n)$
0	$f(0)$	$\Delta(f)(0)$	$\Delta^2(f)(0)$	$\Delta^3(f)(0)$	$\Delta^4(f)(0)$	$\Delta^5(f)(0)$
1	$f(1)$	$\Delta(f)(1)$	$\Delta^2(f)(1)$	$\Delta^3(f)(1)$	$\Delta^4(f)(1)$	$\Delta^5(f)(1)$
2	$f(2)$	$\Delta(f)(2)$	$\Delta^2(f)(2)$	$\Delta^3(f)(2)$	$\Delta^4(f)(2)$	$\Delta^5(f)(2)$
3	$f(3)$	$\Delta(f)(3)$	$\Delta^2(f)(3)$	$\Delta^3(f)(3)$	$\Delta^4(f)(3)$	$\Delta^5(f)(3)$
$\vdots$	$\vdots$	$\vdots$	$\vdots$	$\vdots$	$\vdots$	$\vdots$

The beauty is that you don't even need to have a specific function f in mind to think about this.

Look carefully at this table, and chase $f(3)$ up to the top row using the up and over scheme. You should get:

$$f(3) = \binom{3}{0} f(0) + \binom{3}{1} \Delta(f)(0) + \binom{3}{2} \Delta^2(f)(0) + \binom{3}{3} \Delta^3(f)(0).$$

More generally, we could write:

$$\begin{aligned} f(n) &= \binom{n}{0} f(0) + \binom{n}{1}\Delta(f)(0) + \binom{n}{2}\Delta^2(f)(0) + \binom{n}{3}\Delta^3(f)(0) + \cdots \\ &\quad + \binom{n}{n}\Delta^n(f)(0) \\ &= \sum_{k=0}^{n} \binom{n}{k}\Delta^k(f)(0). \end{aligned}$$

This is a beautiful formula. Some of its features include:

- It tells you how to find the values of a function from the "zeroth" row of its difference table.
- It works for *any* function, not just polynomial functions.
- It works for any nonnegative integer n (with the proviso that $\binom{n}{k} = 0$ if $k > n$).

Example: Let's check out the formula for $f(x) = x^3 + 3x$. Using the definition

$$\Delta(f)(x) = f(x+1) - f(x)$$

we can derive the formulas for the successive differences:

$$\begin{array}{lclcr} f(x) & = & & & x^3 + 3x, \\ \Delta(f)(x)) & = & f(x+1) - f(x) & = & 3x^2 + 3x + 4, \\ \Delta^2(f)(x)) & = & \Delta(f)(x+1) - \Delta(f)(x) & = & 6x + 6, \\ \Delta^3(f)(x)) & = & \Delta^2(f)(x+1) - \Delta^2(f)(x) & = & 6, \\ \Delta^4(f)(x)) & = & \Delta^3(f)(x+1) - \Delta^3(f)(x) & = & 0. \end{array}$$

This allows easy construction of the table:

Notice how the columns go all the way down to the bottom. We have *formulas* now. If the original function is known, so are all its Δs.

	$f(n)$	$\Delta(f)(n)$	$\Delta^2(f)(n)$	$\Delta^3(f)(n)$	$\Delta^4(f)(n)$
n	$n^3 + 3n$	$3n^2 + 3n + 4$	$6n + 6$	6	0
0	0	4	6	6	0
1	4	10	12	6	0
2	14	22	18	6	0
3	36	40	24	6	0
4	76	64	30	6	0
5	140	94	36	6	0
6	234	130	42	6	0
7	364	172	48	6	0

So, let's see:

$$\begin{aligned} f(7) &= \binom{7}{0}0+\binom{7}{1}4+\binom{7}{2}6+\binom{7}{3}6+\binom{7}{4}0+\binom{7}{5}0+\binom{7}{6}0+\binom{7}{7}0 \\ &= 1\cdot 0+7\cdot 4+21\cdot 6+35\cdot 6+35\cdot 0+21\cdot 0+6\cdot 0+1\cdot 0+\cdots \\ &= 364. \end{aligned}$$

Yup. It works. Notice how the Δ columns become 0 before the $\binom{7}{k}$ "run out." Now let's try it with a smaller number:

$$\begin{aligned} f(2) &= \binom{2}{0}0+\binom{2}{1}4+\binom{2}{2}6+\binom{2}{3}6+\binom{2}{4}0+\binom{2}{5}5+\binom{2}{6}0+\cdots \\ &= 1\cdot 0+2\cdot 4+1\cdot 6+0\cdot 6+0\cdot 0+0\cdot 0+0\cdot 0+\cdots \\ &= 14. \end{aligned}$$

Yup again, except this time the $\binom{2}{k}$ became 0 before the Δs ran out.

Finally, notice that for all $n > 2$ (to ensure that $\binom{n}{3} \neq 0$), if we look at our formula algebraically, we recover the formula that defines f:

$$\begin{aligned} f(n) &= \binom{n}{0}0+\binom{n}{1}4+\binom{n}{2}6+\binom{n}{3}6+\binom{n}{4}0+\cdots \\ &= 1\cdot 0+n\cdot 4+\frac{n(n-1)}{2}\cdot 6+\frac{n(n-1)(n-2)}{6}\cdot 6 \\ &\quad +\frac{n(n-1)(n-2)(n-3)}{4!}\cdot 0+\cdots \\ &= n^3+3n. \end{aligned}$$

Again, the proof of this formula is easier to talk about than to write down.

We should state our formula as a theorem and play with it a bit before we move on:

In terms of fitting a function to a table, the theorem says that any function that agrees with a table has to produce values at n that are given by this formula, where the "$\Delta^k(f)(0)$" are taken from the "zeroth" row of the table.

Theorem 3. *If f is any function defined (at least) on the nonnegative integers, then, for all integers $n \geq 0$,*

$$f(n) = \sum_{k=0}^{n} \binom{n}{k} \Delta^k(f)(0).$$

Problems

22. Find a formula for the Δ of each function:

(a) $x \mapsto 4x^2+1$
(b) $x \mapsto 8x^2+2$
(c) $x \mapsto 3x$
(d) $x \mapsto 4x^2+3x+1$
(e) $x \mapsto 3x+5$
(f) $x \mapsto x^3+3x+728$

23. Suppose $f(x) = x^4+5x^3-7x^2+3x-1$.

"Our favorite formula" is the one in Theorem 3.

(a) Find formulas for $\Delta^k(f)$ for $k = 1, 2, 3, 4,$ and 5.
(b) Use part 23a and our favorite formula to find $f(9)$.

24. Here's a difference table:

n	$f(n)$	$\Delta(f)(n)$	$\Delta^2(f)(n)$	$\Delta^3(f)(n)$	$\Delta^4(f)(n)$
0	1	−2	4	6	8
1					8
2					8
3					8
4					8
5					8
6					8
7					8

Assuming $\Delta^4(f)(n) = 8$ for all n, find $f(12)$. Find a formula for $f(n)$.

25. What about functions that are not polynomial? For example, complete the table for $f(n) = 2^n$.

Input	Output	Δ	Δ^2	Δ^3	Δ^4	Δ^5	...
0	1						
1	2						
2	4						
3	8						
4	16						
5							
6							
7							
8							
9							
10							

Express $f(n)$ in terms of the numbers in the first row of the table. Does this lead to anything interesting?

Problems 26–28 let you find the differences of a combinatorial polynomial.

26. Show that if $k \geq 1$, we have a polynomial identity:

$$\binom{x}{k} = \binom{x-1}{k} + \binom{x-1}{k-1}$$

Hint: See problem 20 on page 22.

What happens if $k = 0$?

27. Show that, for $k \geq 1$,

$$\Delta\left(\binom{x}{k+1}\right) = \binom{x}{k}$$

28. State and prove a theorem that gives you a formula for

$$\Delta^n \left(x \mapsto \binom{x}{m} \right)$$

for any nonnegative integers m and n.

Problems 29–31 explore the differences of a power of x.

29. Find formulas for the Δ of each function.

(a) $x \mapsto 1$

(b) $x \mapsto x$

(c) $x \mapsto x^2$

(d) $x \mapsto x^3$

(e) $x \mapsto x^4$

(f) $x \mapsto 5x^4 - 2x^3 + 3x^2 - x - 1$

Hint: Use the Binomial Theorem.

30. Show that

$$\Delta(x \mapsto x^n)$$

is a polynomial of degree $n - 1$. What's the degree of $\Delta^2(x \mapsto x^n)$?

31. Show that, if $k > n$, $\Delta^k(x \mapsto x^n) = 0$.

Is this really an "infinite" sum?

32. Show that, if f is a polynomial function, f can be written as

$$f(x) = \sum_{k=0}^{\infty} \Delta^k(f)(0) \binom{x}{k}.$$

33. Let $f(x) = 5x^6 + 4x^5 + 3x^4 - 2x^3 + x^2 - x - 8$. For integers $r \geq 1$, let

$$f_r(x) = \sum_{k=0}^{r} \Delta^k(f)(0) \binom{x}{k}.$$

What would f_8 be?

(a) Tabulate each of the functions $f, f_1, f_2, f_3, f_4, f_5, f_6$, and f_7 between 0 and 10. How are the tables related?

(b) Sketch the graphs of $f, f_1, f_2, f_3, f_4, f_5, f_6$, and f_7. How are they related?

34. Same as problem 33 except $f(x) = 2^x$.

35. If f is any function and r is an integer, $r \geq 1$, let

$$f_r(x) = \sum_{k=0}^{r} \Delta^k(f)(0) \binom{x}{k}.$$

Show that $\Delta^k(f)(0) = \Delta^k(f_r)(0)$ if $0 \leq k \leq r$.

36. In what sense is the result of problem 32 true if $f(x) = 2^x$?

37. Theorem 2 allows you to write any function whose differences are eventually constant as a *linear combination* (that is, as a sum of constant multiples) of the $\binom{x}{k}$. Write each polynomial function as a linear combination of the $\binom{x}{k}$:

(a) $x \mapsto 1$ **(b)** $x \mapsto x$ **(c)** $x \mapsto x^2$

(d) $x \mapsto x^3$ **(e)** $x \mapsto x^4$ **(f)** $x \mapsto x^5$

Problems 38–41 look at differences for some nonpolynomial functions.

38. Suppose $f(n) = 3^n$.

(a) Describe the difference table for f. In particular, what is a formula for $\Delta^k(f)(0)$?

(b) Express $f(n)$ in terms of the $\Delta^k(f)(0)$.

39. Suppose $f(n) = a^n$.

(a) Describe the difference table for f. In particular, what is a formula for $\Delta^k(f)(0)$?

(b) Express $f(n)$ in terms of the $\Delta^k(f)(0)$.

40. Suppose $f(n) = (1 + a)^n$.

(a) Describe the difference table for f. In particular, what is a formula for $\Delta^k(f)(0)$?

(b) Express $f(n)$ in terms of the $\Delta^k(f)(0)$.

41. Suppose $f(n)$ = the nth Fibonacci number. So, the table for f looks like this:

Any output (after 0) is the sum of the previous two outputs.

n	$f(n)$
0	0
1	1
2	1
3	2
4	3
5	5
6	8
7	13

(a) Describe the difference table for f. In particular, find a formula for $\Delta^k(f)(0)$ in terms of the f column.

(b) Express $f(n)$ in terms of the $\Delta^k(f)(0)$.

A closed form for this table is particularly difficult. Look for ways to relate the differences and the $f(n)$.

42. Suppose you have a mystery function g with the property that $\Delta(g)(x) = 5x - 2$. Find formulas for:

(a) $\Delta\big(x \mapsto 2\,g(x)\big)$ **(b)** $\Delta\big(x \mapsto 2\,g(x) + x^3\big)$

(c) $\Delta\big(x \mapsto 2\,g(x) + 7\big)$ **(d)** $\Delta\big(x \mapsto 2\,g(x) + 9\big)$

What would you need to know to find a formula for $g(x)$?

43. Suppose $f(x)$ is a polynomial. Show that f is a constant function if and only if $\Delta(f) = 0$.

44. Suppose $f(x)$ is a polynomial. Show that you can construct $\Delta(f)$ by the following process:

(a) Write f as a sum of powers of x (that is, remove all parentheses and write f in "normal" form).

(b) Replace each power x^k in f by $(x+1)^k - x^k$.

(c) Simplify the result.

Problems 45 and 46 are not necessary for what follows. But they are fun.

45. Using the ideas from this section, write a convincing argument to show that, if f is a polynomial and $\Delta^3(f)$ is constant and nonzero, f has degree 3.

46. Using the ideas from this section, write a convincing argument to show that, if f is a polynomial and $\Delta^m(f)$ is constant and nonzero, f has degree m.

Historical Perspective

The expansion of a polynomial f in problem 32:

$$f(x) = \sum_{k=0}^{\infty} \Delta^k(f)(0)\binom{x}{k}$$

actually holds for a much wider class of functions than polynomials (although the sum is no longer finite for nonpolynomials). Some people call this expression the "Mahler expansion" for f, and they even call the polynomials $\binom{x}{k}$ the "Mahler polynomials." This has nothing to do with the composer Gustav Mahler; rather it is in honor of the German-born mathematician Kurt Mahler (1903–1988). In the 1960s, Mahler showed that this kind of expansion could be used to extend some important classical functions to a number system useful in number theory (and known as the "p-adic numbers").

1.5 Going the Other Way: Polynomials to Tables

So, if we assume (or even better, know) that the mth difference in a table is constant and nonzero, we know that there is a polynomial of degree m that agrees with the table. In fact, that polynomial can be expressed as:

$$\sum_{k=0}^{m} a_k \binom{x}{k}$$

where the a_k are the entries in the zeroth row of the difference table (before they all become 0). There's one open question: If you start with a polynomial of degree m, will its mth differences be constant?

The answer is yes. To see this, we'll need two things:

- We'll prove an important property of the Δ operator known as *linearity*. Linearity will allow us to take the Δ of a linear combination of polynomials by taking the Δ of each polynomial in the sum.

- We'll need the following result:

$$\text{If } k > n,\ \Delta^k(x \mapsto x^n) = 0.$$

Since any polynomial of degree m is a linear combination of the various x^n, these two results will allow us to repeatedly take the Δ of a polynomial until we get 0. Counting how many times this takes will give us the result we want.

First, let's handle the linearity. For the sake of example, let's say that

By "$f+g$," we mean the sum of the polynomials f and g. You can think of this as the sum of two functions or as the high school algebra style sum of two polynomial expressions. Either way, $f+g$ is the thing you get by "adding and combining like terms." Same for $c\,f$: multiply each coefficient in f by c.

$$f(x) = x^2 + 1,$$
$$g(x) = x^3 - 2x,$$

so

$$(f+g)(x) = x^3 + x^2 - 2x + 1$$

and

$$(3f)(x) = 3x^2 + 3.$$

Here's a first difference table for f and g, their sum $f+g$, and $3f$:

n	$f(n)$	$\Delta(f)(n)$	$g(n)$	$\Delta(g)(n)$	$(f+g)(n)$	$\Delta(f+g)(n)$	$(3f)(n)$	$\Delta(3f)(n)$
0	1	1	0	−1	1	0	3	3
1	2	3	−1	5	1	8	6	9
2	5	5	4	17	9	22	15	15
3	10	7	21	35	31	42	30	21
4	17	9	56	59	73	68	51	27
5	26	11	115	89	141	100	78	33

Notice that you can get the column for $\Delta(f+g)$ in two ways:

So, "the difference of the sum is the sum of the differences."

- You can take the differences for the $f+g$ column, or
- You can add the $\Delta(f)$ and $\Delta(g)$ columns.

So, "the difference of the multiple is the multiple of the difference."

And, you can get the column for $\Delta(3f)$ in two ways:

- You can take the differences for the $3f$ column, or
- You can multiply the $\Delta(f)$ column by 3.

This works in general, and is worthy of being called a theorem:

Caution: the first bullet is not the distributive law. The left-hand side is the difference of the sum, and the right-hand side is the sum of the differences. Similarly, the second bullet is not an algebraic rearrangement. Can you say it in words?

Theorem 4. (The linearity of Δ.) *Suppose f and g are polynomials and c is a number. Then*

- $\Delta(f+g) = \Delta(f) + \Delta(g)$,
- $\Delta(cf) = c\Delta(f)$.

Proof Let's first work out an example that is "general in principle," an example that is sort of halfway between a numerical example and a full-blown proof and that contains all the essential ingredients of the proof.

Suppose $f(x) = a_3x^3 + a_2x^2 + a_1x + a_0$ and $g(x) = b_2x^2 + b_1x + b_0$ where the a_i and b_j are numbers.

First, let's compute $\Delta(f) + \Delta(g)$.

$$\begin{aligned}\Delta(f)(x) &= f(x+1) - f(x)\\ &= a_3(x+1)^3 + a_2(x+1)^2 + a_1(x+1)\\ &\quad + a_0 - (a_3x^3 + a_2x^2 + a_1x + a_0)\\ &= a_3\big((x+1)^3 - x^3\big) + a_2\big((x+1)^2 - x^2\big) + a_1\big((x+1) - x\big).\end{aligned}$$

Similarly, $\Delta(g)(x) = b_2\big((x+1)^2 - x^2\big) + b_1\big((x+1) - x\big)$, so

$$\begin{aligned}\big(\Delta(f) + \Delta(g)\big)(x) &= a_3\big((x+1)^3 - x^3\big) + (a_2 + b_2)\big((x+1)^2 - x^2\big)\\ &\quad + (a_1 + b_1)\big((x+1) - x\big).\end{aligned}$$

But if you look at

$$\begin{aligned}(f+g)(x) &= f(x) + g(x)\\ &= a_3x^3 + (a_2 + b_2)x^2 + (a_1 + b_1)x + (a_0 + b_0)\end{aligned}$$

and take its Δ, you'll end up with the same thing. Try it.

Proofs like this are just descriptions of generic calculations. They are easier to follow if you sketch out some details of an *actual* calculation as you read them.

The proof in general is essentially the same. To get the Δ of an arbitrary polynomial f, you look at $f(x)$ and replace each x^k by $(x+1)^k - x^k$. You do the same to get the Δ of g. But then adding $\Delta(f)$ and $\Delta(g)$ is just like adding f and g, except each occurrence of x^k is replaced by $(x+1)^k - x^k$. And this is exactly how you form $\Delta(f+g)$.

The proof that $\Delta(cf) = c\Delta(f)$ follows similar reasoning, and is left as a problem for you. ■

More generally, since $\Delta^k(f) = \Delta\big(\Delta^{k-1}(f)\big)$ (and since $\Delta^{k-1}(f)$ is a function in its own right), we can "argue down" to the following generalization (which you will prove in problem 51):

Theorem 5. *If $k \geq 1$ is an integer, the operator Δ^k is linear in the sense of Theorem 4:*

- $\Delta^k(f+g) = \Delta^k(f) + \Delta^k(g)$,
- $\Delta^k(cf) = c\Delta^k(f)$.

The theorem is often described by the saying "The Δ of a sum is the sum of the Δs, and numbers move out."

This theorem will help a great deal in proving that the mth difference in a polynomial of degree m is constant. Suppose, for example, we want to find the third difference of the polynomial $f(x) = 5x^3 - 3x^2 + 2x - 1$. You could calculate like this:

$$\begin{aligned}\Delta^3(f)(x) &= \Delta^3(5x^3 - 3x^2 + 2x - 1)\\ &= \Delta^3(5x^3) + \Delta^3(-3x^2) + \Delta^3(2x) + \Delta^3(-1)\\ &= 5\Delta^3(x^3) - 3\Delta^3(x^2) + 2\Delta^3(x) - \Delta^3(1).\end{aligned}$$

So, once we know how to find the Δ^3 of a "pure" power x^n, we're done. And the next theorem tells us how to do that:

Theorem 6. *If k and n are nonnegative integers, then*

1. *$\Delta^k(x^n)$ has degree $n - k$ if $k \le n$,*
2. $\Delta^n(x^n) = n!$,
3. *$\Delta^k(x^n) = 0$ if $k > n$.*

Proof $\Delta(x^n) = (x+1)^n - x^n$. But, by the binomial theorem,

The binomial theorem:

$$(a+b)^n = \sum_{k=0}^{n} \binom{n}{k} a^{n-k} b^k$$

For a proof, see [15].

$$\begin{aligned}(x+1)^n &= \sum_{k=0}^{n} \binom{n}{k} x^k\\ &= x^n + \binom{n}{n-1}x^{n-1} + \binom{n}{n-2}x^{n-2} + \cdots + \binom{n}{0}.\end{aligned}$$

So,

$$\Delta(x^n) = \binom{n}{n-1}x^{n-1} + \binom{n}{n-2}x^{n-2} + \cdots + \binom{n}{0},$$

a polynomial of degree $n - 1$. Now proceed inductively:

$$\begin{aligned}\Delta^2(x^n) &= \Delta(\Delta(x^n))\\ &= \Delta\left(\binom{n}{n-1}x^{n-1} + \binom{n}{n-2}x^{n-2} + \cdots + \binom{n}{0}\right)\\ &= \binom{n}{n-1}\Delta(x^{n-1}) + \binom{n}{n-2}\Delta(x^{n-2}) + \cdots + \binom{n}{0}\Delta(1)\\ &= \binom{n}{n-1}(\text{a polynomial of degree } n-2)\\ &\quad + \binom{n}{n-2}(\text{a polynomial of degree } n-3) + \cdots\\ &= \text{a polynomial of degree } n-2.\end{aligned}$$

The same argument works for Δ^3 and, more generally, for Δ^k.

To get the second statement of the theorem, notice that, by the first statement, $\Delta^n(x^n)$ has degree $n - n = 0$. A polynomial of degree 0 is a constant. A little more work will get the exact values of this constant (see problem 52). And, once we get to a constant, the next (and subsequent) Δs are all 0. ■

Well, we're now in a position to prove the main result of this section, the converse to Theorem 2:

Theorem 7. *If f is a polynomial of degree m, $\Delta^m(f)$ is constant.*

Proof Imagine taking Δ^m of f. By linearity, you could take Δ^m of each term in f. But all the terms except one have degree less than m, so, by the last part of Theorem 6, each such term would vanish when you take Δ^m of it. That leaves the highest degree term, which is a number times x^m. But Δ^m of that is a constant, by the second part of Theorem 6. ■

Problems

47. Show that if f is a polynomial of degree m, $\Delta(f)$ is a polynomial of degree $m-1$.

48. Suppose f is polynomial and $g(x) = f(x) + c$, where c is some number. Show that $\Delta f = \Delta g$.

49. Suppose f and g are two polynomials and $\Delta f = \Delta g$. What can you conclude about f and g?

50. Prove that $\Delta(cf) = c\Delta(f)$.

51. Prove Theorem 5.

52. Show that $\Delta^n(x^n) = n!$.

53. Theorem 7 says that the mth difference of a polynomial of degree m is a number. What *is* that number, in terms of the original polynomial?

Try some examples and look for patterns.

54. Find a closed form for the sum

$$\sum_{k=0}^{n}(-1)^k k^n \binom{n}{k}.$$

55. Show that if f is *any* function, polynomial or not, so that a polynomial of degree m agrees with $\Delta(f)$ on positive integers, then there is a polynomial of degree $m+1$ that agrees with f on positive integers.

Problems 56–59 show how to express the kth difference for a function in terms of the function's outputs.

Here, x can be any input to f. If you replace x by 3, you get $\Delta(f)(3) = f(4) - f(3)$. If you replace x by $x+1$, you get $\Delta(f)(x+1) = f(x+2) - f(x+1)$.

This is subtle. It might help to replace $\Delta(f)$ in the first two lines by some temporary symbol like g.

56. No matter what function f we have, we know that

$$\Delta(f)(x) = f(x+1) - f(x).$$

Well, $\Delta^2(f) = \Delta\big(\Delta(f)\big)$, so, we can apply the above formula to $\Delta(f)$ and obtain a formula for $\Delta^2(f)$:

$$\begin{aligned}\Delta^2(f)(x) &= \Delta\big(\Delta(f)\big)(x)\\ &= \Delta(f)(x+1) - \Delta(f)(x)\\ &= \big(f(x+2) - f(x+1)\big) - \big((f(x+1) - f(x)\big)\\ &= f(x+2) - 2f(x+1) + f(x).\end{aligned}$$

(a) Supply a reason for each step in this calculation.

(b) Use this result to find a formula for $\Delta^2(f)$ where $f(x) = 5x^2 - 3x + 2$.

57. Show that, for any function f,

$$\Delta^3(f)(x) = f(x+3) - 3f(x+2) + 3f(x+1) - f(x).$$

58. More generally, explain why the following theorem is true:

This theorem will be useful in the next section. To get started, write out the right-hand side without the Σ notation. Then use the rhythm of problems 56 and 57.

Theorem 8. *Suppose f is any function and k is a nonnegative integer. Then*

$$\Delta^k(f)(x) = \sum_{j=0}^{k} (-1)^j \binom{k}{j} f(x+k-j).$$

59. Suppose m is any positive integer and $f(x) = x^m$. Suppose further that k is some positive integer. Show that

Try special cases. Let $m = 4$ and $k = 3$, for example.

$$\Delta^k(f)(0) = a_{m,k} = k^m - \binom{k}{1}(k-1)^m + \binom{k}{2}(k-2)^m - \cdots \pm \binom{k}{k-1} 1^m.$$

Use this formula to get the "zeroth" row of the difference table for $f(x) = x^5$.

60. Notice that

$$4^3 - \binom{4}{1}3^3 + \binom{4}{2}2^3 - \binom{4}{3}1^3 = 0.$$

Show, more generally, that if $k > m$,

Hint. Use problems 59 (above) and 31 (on page 28).

$$k^m - \binom{k}{1}(k-1)^m + \binom{k}{2}(k-2)^m - \cdots \pm \binom{k}{k-1} 1^m = 0.$$

1.6 Conversions

We now have two useful ways of writing polynomials:

The powers of x are sometimes called the "standard basis" for polynomials. The $\binom{x}{k}$ is sometimes called the "Mahler basis."

- in "standard form" as sums of monomials, just as in first-year algebra. You can think of such an expression as a "linear combination" of powers of x. For example, $x^3 - x^2 + 3x - 4$ is just a sum of powers of x, with number coefficients (4 is $4x^0$).
- as a linear combination of the polynomials $\binom{x}{k}$. For example, $x^3 - x^2 + 3x - 4$, when written in this form, looks like

$$6\binom{x}{3} + 4\binom{x}{2} + 3\binom{x}{1} - 4\binom{x}{0}.$$

The next problems look into converting between one form and the other. It starts by looking at how you express the "atoms" of one basis—a single power of x or a single combinatorial polynomial—in terms of the other. The coefficients in these simple conversions turn out to be quite important; we'll meet them again later on. The later problems look at more general conversions.

Problems

Hint: Use problem 59 on page 35 and Theorem 2 on page 18. Calculate a few of the $a_{m,k}$ (for, say, $m = 0$ to 6), just to get familiar with them. They'll make a surprise appearance later in Chapter 4.

To make this work, we need to take the "ambiguous" 0^0 to be 1. Why?

61. Show that, if m is a positive integer,

$$x^m = \sum_{k=0}^{m} a_{m,k} \binom{x}{k}$$

where

$$a_{m,k} = k^m - \binom{k}{1}(k-1)^m + \binom{k}{2}(k-2)^m - \cdots + (-1)^{k-1}\binom{k}{k-1}1^m.$$

62. Problem 61 shows how to write powers of x in terms of the combinatorial polynomials. Can you do it the other way? If $\binom{x}{m}$ is expanded as powers of x, what can you say about the coefficients? For example,

$$\binom{x}{4} = \frac{x(x-1)(x-2)(x-3)}{4!}$$
$$= \frac{x^4 - 6x^3 + 11x^2 - 6x}{4!}.$$

In general, the signs will alternate and the denominator of $\binom{x}{m}$ will be $m!$.

Define the numbers $b_{m,k}$ by

$$\binom{x}{m} = \frac{x(x-1)(x-2)(x-3)\cdots(x-m+1)}{m!} \qquad \text{(now expand this)}$$
$$= \sum_{k=0}^{m} b_{m,k} x^k.$$

What patterns can you find among the $b_{m,k}$?

Try writing out the results of problems 63 and 64 in an array, Pascal triangle style. You'll pick up the rhythm quickly.

63. Using the notation of problem 61, show that the $a_{m,k}$ can be defined recursively by:

$$a_{m,k} = \begin{cases} 1 & \text{if } m = k = 0, \\ 0 & \text{if } k > m \text{ or } k < 0, \\ k(a_{m-1,k-1} + a_{m-1,k}) & \text{if } 0 \le k \le m. \end{cases}$$

64. Using the notation of problem 62, show that the $b_{m,k}$ can be defined recursively by:

$$b_{m,k} = \begin{cases} 1 & \text{if } m = k = 0, \\ 0 & \text{if } k > m \text{ or } k < 0, \\ \dfrac{b_{m-1,k-1} - (m-1)b_{m-1,k}}{m} & \text{if } 0 \le k \le m. \end{cases}$$

65. Let

$$f(x) = 3x^5 - 2x^4 + 5x^3 - 4x + 7.$$

Write f as a linear combination of the $\binom{x}{k}$.

66. Let

$$f(x) = 3\binom{x}{5} - 2\binom{x}{4} + 5\binom{x}{3} - 4\binom{x}{1} + 7\binom{x}{0}.$$

Write f as a linear combination of the powers of x.

Would a CAS change the way you answer these last two problems?

Problems 67–70 show how to "interpret" the coefficients of a polynomial when it is written in terms of the Mahler basis.

67. Show that for any polynomial f there's only one way to write it as a sum of powers of x.

68. Suppose a is a number. Show that for any polynomial f there's only one way to write it as a sum of powers of $x - a$.

69. Show that for any polynomial f there's one and only one way to write it as a sum of combinatorial polynomials $\binom{x}{k}$.

70. Prove Theorem 9 below.

Theorem 9. *If f is a polynomial of degree m,*

$$f(x) = \sum_{k=0}^{m} \Delta^k(f)(0)\binom{x}{k}.$$

This refines the result of problem 32 on page 28. It shows how to "interpret" the coefficients a_k when you write a polynomial of degree m in the form $\sum_{k=0}^{m} a_k\binom{x}{k}$.

71. Show that if f is a polynomial of degree m,

$$\Delta(f)(x) = \sum_{k=1}^{m} \Delta^k(f)(0)\binom{x}{k-1}.$$

72. Consider the sequence of polynomials f_m where

$$f_m(x) = \sum_{k=0}^{m} a_k\binom{x}{k+1}$$

and

$$a_k = k^m - \binom{k}{1}(k-1)^m + \binom{k}{2}(k-2)^m - \cdots + (-1)^{k-1}\binom{k}{k-1}1^m.$$

Find a simple formula for $\Delta(f_m)$.

It's a good idea to calculate the first few f_m by hand: f_1, f_2, f_3, and f_4. Then calculate the Δ of each.

73. Show that if f is a polynomial and $\Delta(f)$ has degree m, then f has degree $m + 1$.

74. Show that, if m is a positive integer,

$$\Delta(x^m) = \sum_{k=1}^{m} a_k\binom{x}{k-1}$$

where

$$a_k = k^m - \binom{k}{1}(k-1)^m + \binom{k}{2}(k-2)^m - \cdots + (-1)^{k-1}\binom{k}{k-1}1^m.$$

1.7 From Newton to Lagrange

Newton's difference formula (Theorem 2) gives us a nice way to fit a polynomial function to a table. It doesn't work, however, unless the inputs are consecutive integers, starting at 0.

Suppose you have a random input-output table with no pattern to the inputs. Maybe you got the table through some data-gathering device or through an empirical experiment. At any rate, you want to fit a polynomial to the data.

This method is connected to a famous theorem in number theory, the "Chinese remainder theorem."

In this section, we'll look at a method for doing this called *Lagrange interpolation.* If you have access to a CAS, it provides a method as simple as Newton's, and it's even more general.

Here's a table:

Input	Output
1	−6
3	166
6	7159
8	31291
9	56938
12	243787

We'll find a polynomial f of degree at most 5 that agrees with the table. First of all, here's a polynomial that will be *zero* at each of the inputs on the table:

$$(x-1)(x-3)(x-6)(x-8)(x-9)(x-12).$$

And, if we remove any one of the factors, the resulting expression will vanish at *all but one* of the inputs. Lagrange's idea is to capitalize on this and to write a polynomial in a form that is easy to evaluate at these particular inputs. He considers:

$$\begin{aligned} f(x) = \quad & A(x-1)(x-3)(x-6)(x-8)(x-9) \\ & + B(x-1)(x-3)(x-6)(x-8)(x-12) \\ & + C(x-1)(x-3)(x-6)(x-9)(x-12) \\ & + D(x-1)(x-3)(x-8)(x-9)(x-12) \\ & + E(x-1)(x-6)(x-8)(x-9)(x-12) \\ & + F(x-3)(x-6)(x-8)(x-9)(x-12). \qquad (*) \end{aligned}$$

The numbers A–F will be determined in a minute. Each product is formed by taking

$$(x-1)(x-3)(x-6)(x-8)(x-9)(x-12)$$

and "dropping" one factor. So, each product is a polynomial of degree 5, and *that* ensures that the whole expression will be of degree at most 5.

Why write f in such a messy way? Well, it allows you to easily calculate $f(n)$ for any n between 0 and 5. For example, from the table, we want $f(1)$ to be -6. So, we calculate like this:

$$\begin{aligned} f(1) = -6 = \quad & A(1-1)(1-3)(1-6)(1-8)(1-9) \\ & + B(1-1)(1-3)(1-6)(1-8)(1-12) \\ & + C(1-1)(1-3)(1-6)(1-9)(1-12) \\ & + D(1-1)(1-3)(1-8)(1-9)(1-12) \\ & + E(1-1)(1-6)(1-8)(1-9)(1-12) \\ & + F(1-3)(1-6)(1-8)(1-9)(1-12). \end{aligned}$$

But look: all the terms except the last have a factor of 0, so they all vanish. We get

$$-6 = F(-2)(-5)(-7)(-8)(-11) = -6160F.$$

So,

$$F = \frac{-6}{-6160} = \frac{3}{3080}.$$

Next, let $x = 3$. From the table, we want $f(3)$ to be 166. But when we replace x by 3 in expression $(*)$, the only term to survive is the "E" term, and we get:

$$166 = E(2)(-3)(-5)(-6)(-9) = 1620E.$$

So

$$E = \frac{83}{810}.$$

Similarly, we can pick off the other missing coefficients by replacing x by 2 (producing $D = -7159/540$), 8 (producing $C = 31291/280$), 9 (producing $B = -28469/216$), and 12 (producing $A = 243787/7128$). So, substituting in $(*)$, we get:

$$\begin{aligned} f(x) = \quad & \frac{243787}{7128}(x-1)(x-3)(x-6)(x-8)(x-9) \\ & - \frac{28469}{216}(x-1)(x-3)(x-6)(x-8)(x-12) \\ & + \frac{31291}{280}(x-1)(x-3)(x-6)(x-9)(x-12) \end{aligned}$$

$$- \frac{7159}{540}(x-1)(x-3)(x-8)(x-9)(x-12)$$
$$+ \frac{83}{810}(x-1)(x-6)(x-8)(x-9)(x-12)$$
$$+ \frac{3}{3080}(x-3)(x-6)(x-8)(x-9)(x-12).$$

A CAS simplifies this to $x^5 - 3x^3 + x^2 - 5$. Try it for yourself.

Facts and Notation

In the following problem set, you'll be asked to compare and contrast Newton's difference formula and Lagrange interpolation. When doing so, it's useful to use some results from algebra. The proofs of most of the following can be found in [15]. If you've taught second-year algebra, you probably know these quite well. Here's our little algebra refresher:

And we'll sketch the proofs in the next chapter.

By "polynomial" here we mean "polynomial in one variable with real coefficients."

- **Long division.** If f and g are polynomials, there exist polynomials q and r such that the degree of r is strictly less than the degree of g and

$$f = gq + r$$

We'll look at the proofs of some of the other results stated here later (in Chapter 2).

- **The remainder theorem.** If f is a polynomial, then the remainder when $f(x)$ is divided by $x - a$ is $f(a)$.

  The proof is too nice and too short to leave out: Divide $f(x)$ by $x - a$ to get a quotient q and a remainder r. Since the degree of $x - a$ is 1, the degree of r is 0; that is, r is a *number* (independent of x):

$$f(x) = (x-a)q(x) + r.$$

  Replace x by a. You get

$$f(a) = (a-a)q(a) + r = r.$$

Corollary: A polynomial with more roots than its degree is identically 0.

- **The factor theorem.** If f is a polynomial, $(x - a)$ is a factor of $f(x)$ if and only if $f(a) = 0$.
- **The number of roots theorem.** A polynomial of degree n can have at most n roots.
- **The function implies form theorem.** If two polynomials of degree at most n agree for more than n inputs, they are exactly the same polynomial.

Problems

75. Use the first two facts above to prove the last three.

76. Here's a table:

Input	Output
0	3
1	4
2	7
3	48
4	211

(a) Before you pick up a pencil (or a CAS), predict the degrees of the polynomials that Newton and Lagrange will produce.

(b) Use Newton to find a polynomial that agrees with the table.

(c) Use Lagrange to find a polynomial that agrees with the table.

(d) Which method is easier for you?

(e) Do the two methods give you the same result?

Which method do you like better?

77. A radio show offered a prize to the first caller who could predict the next term in this sequence:

$$\{1, 2, 4, 8, 16\}$$

(a) What would you get if you used "common sense?"

(b) What would you get if you used Newton?

(c) What would you get if you used Lagrange interpolation?

78. Another radio show offered a prize to the first caller who could predict the next term in this sequence:

$$\{14, 3, 26, 8, 30\}$$

After no one got it for a few days, the host announced that these are the first five numbers that were retired from the Mudville Sluggers baseball team.

(a) Use Newton to predict the next number that was retired in Mudville.

(b) Use Lagrange to predict the next number that was retired in Mudville.

79. If a table has n inputs (say, 0 to $n-1$), what's the degree of the polynomial fit produced by Newton? By Lagrange?

80. Suppose you have a table that you know was generated by a quadratic, say $f(x) = x^2 - 3x + 5$, but you generate outputs between 0 and 4:

Input	Output
0	5
1	3
2	3
3	5
4	9

It seems that this table should fool our two methods, producing a degree 4 polynomial. Does it? Work it out for both Newton and Lagrange and see what happens.

81. Find a polynomial that agrees with this table:

Input	Output
4	$-\frac{1}{30}$
5	$-\frac{1}{12}$
6	$-\frac{1}{6}$
7	$-\frac{7}{24}$
8	$-\frac{7}{15}$
9	$-\frac{7}{10}$
10	-1

Hint: If a table has inputs 0 through 5, what's the maximum degree of the polynomial that each method produces?

82. Show that Lagrange and Newton produce exactly the same polynomial.

83. Show that our two methods (which produce the same thing by problem 82) produce the polynomial of *lowest degree* that agrees with a given table.

84. Show that two polynomials agree at a set of inputs if and only if their difference vanishes at those inputs.

1.8 Agreeing to Disagree

Those problems on standardized tests that ask you to find the next term in a sequence are the butt of a host of jokes among mathematicians. The jokes all revolve around the fact that there are many functions that agree with a table, so if the test-maker gives you the sequence $1, 2, 4, 8, \cdots$ and asks for the next term, you can find a simple polynomial function that agrees with this table:

Input	Output
0	1
1	2
2	4
3	8
4	1076

Let's take a look at the class of *all* polynomials that agree with a table.

While Lagrange and Newton give you the polynomial of lowest degree that agrees with a table, there are many different polynomials of higher degrees that will work. For example, in problem 80, you used one of our methods to produce a polynomial fit to this table:

Input	Output
0	3
1	4
2	7
3	48
4	211

If you got what we did, you got $f(x) = 2x^4 - 6x^3 + 5x^2 + 3$. But here are some other polynomials that work:

- $g(x) = x^5 - 8x^4 + 29x^3 - 45x^2 + 24x + 3$
- $h(x) = 2x^5 - 18x^4 + 64x^3 - 95x^2 + 48x + 3$
- $j(x) = x^6 - 15x^5 87x^4 - 231x^3 + 279x^2 - 120x + 3$

How'd we get these? We used the following trick:

- $g(x) = f(x) + x(x-1)(x-2)(x-3)(x-4)$
- $h(x) = f(x) + 2x(x-1)(x-2)(x-3)(x-4)$
- $j(x) = f(x) + (x-5)x(x-1)(x-2)(x-3)(x-4)$

In other words, any polynomial of the form

Well, numbers are polynomials, but you know what we mean.

$$f(x) + \lambda x(x-1)(x-2)(x-3)(x-4)$$

where λ is *any* number or polynomial will agree with f at 0–4.

And guess what? The converse is true, too.

Theorem 10. *Two polynomials agree at*

$$\{0, 1, 2, 3, \dots, n\}$$

if and only if they differ by a (polynomial) multiple of

$$x(x-1)(x-2)(x-3)\cdots(x-n).$$

Proof Suppose f and g agree on $0, \dots, n$. Then $f - g = 0$ on these numbers. Since $f - g$ vanishes at 0, the factor theorem implies that

See problem 84 on page 42.

$$f(x) - g(x) = xh(x)$$

for some polynomial h. But then $h(x)$ vanishes at 1 (why?), so by the factor theorem again,

$$xh(x) = x(x-1)h_1(x)$$

for some polynomial h_1. But then $h_1(x)$ vanishes at 2 (why?), so by the factor theorem again,

$$x(x-1)h_1(x) = x(x-1)(x-2)h_2(x)$$

for some polynomial h_2.

You get the picture. Keep up the good work and you end up with

$$f(x) - g(x) = x(x-1)(x-2)\cdots(x-n)h_n(x)$$

So $f(x) - g(x)$ is a multiple of the desired product. ■

Problems

85. Find a polynomial of degree 4 that agrees with $x^3 - 3x^2 + 3x - 1$ at 0, 1, 2, and 3.

86. Find a polynomial that agrees with $x \mapsto x^3 - 3x^2 + 3x - 1$ at 0, 1, 2, and 3 but takes on the value 7 at 5.

87. Nina was taking a test that asked to fill in the next term of the sequence

$$\{1, 2, 4, 8 \ldots\}.$$

Nina answered "761." "How'd you get that?" asked the teacher. Nina produced a polynomial function that tabulated like this:

Input	Output
0	1
1	2
2	4
3	8
4	761

What function might Nina have used?

88. Find a polynomial function g that agrees with $f(x) = 5x + 7$ *only* at 1, 2, 3, and 4. Sketch the graphs of f and g on the same axes.

89. Show that every polynomial function that vanishes at 0–4 is a multiple of $\binom{x}{5}$. Generalize.

90. Suppose f and g are two polynomials that vanish on the set of integers between 0 and n (any set of numbers will do). Show that the following polynomials also vanish on this set:

- $f + g$
- $f - g$
- $(x^2 + 1)f - (2x + 3)g$

Notes for Selected Problems

Notes for problem 2 on page 3. Here is a "proof by example." Using the "up and over" property, 125 is $49+76$. But 49 is $11+38$. Continuing in this manner, we obtain

Refer to the table from problem 2.

$$\begin{aligned}125 &= 49+76\\ &= (11+38)+76\\ &= (-1+12)+38+76\\ &= 1+(-2)+12+38+76.\end{aligned}$$

Notes for problem 4 on page 8. Here, we can apply the hockey-stick property:

Input	Output	Δ
0	c	d
1		d
2		d
3		d
4		d
5		

$$f(5) = c + (d+d+d+d+d) = 5d + c.$$

Of course, in general, we have $f(n) = dn + c$. Thus, a constant Δ column means that a linear polynomial fits the table.

Notes for problem 7 on page 9. We get a formula that agrees with the Δ column by starting with b and stacking up cs:

$$\Delta = b + cn.$$

Then, to get the output at n, you start with a and stack up n Δs, so the output at n is

$$\begin{aligned}&a + (b+0c) + (b+c) + (b+2c) + \cdots + \big(b+(n-1)c\big)\\ &= a + nb + c(1+2+\cdots+n-1)\\ &= a + nb + c\frac{n(n-1)}{2}\\ &= a + \Big(b - \frac{c}{2}\Big) + \frac{c}{2}n^2.\end{aligned}$$

Notes for problem 8 on page 9. As shown below, the third differences are constant. If we cover up the $f(n)$ column, we see that the second differences of the Δ column (i.e., the Δ^3 column) are constant. So, using our result from problem 7 with $a=-2$, $b=14$, and $c=12$, we get $\Delta = 6n^2 + 8n - 2$.

n	$f(n)$	Δ	Δ^2	Δ^3
0	1	−2	14	12
1	−1	12	26	12
2	11	38	38	12
3	49	76	50	12
4	125	126	62	12
5	251	188	74	
6	439	262		
7	701			

Now, we rewrite the table as follows:

n	$f(n)$	Δ
0	1	$-2 = 6 \cdot 0^2 + 8 \cdot 0 - 2$
1	−1	$12 = 6 \cdot 1^2 + 8 \cdot 1 - 2$
2	11	$38 = 6 \cdot 2^2 + 8 \cdot 2 - 2$
3	49	$76 = 6 \cdot 3^2 + 8 \cdot 3 - 2$
4	125	$126 = 6 \cdot 4^2 + 8 \cdot 4 - 2$
5	251	$188 = 6 \cdot 5^2 + 8 \cdot 5 - 2$
6	439	$262 = 6 \cdot 6^2 + 8 \cdot 6 - 2$
7	701	

Equipped with this new table and using the hockey-stick property, you should be able to show that

$$f(4) = 1 + 6(0^2 + 1^2 + 2^2 + 3^2) + 8(0 + 1 + 2 + 3) + 4(-2).$$

And, in general, we have:

$$\begin{aligned} f(n) &= 1 + 6\big(0^2 + 1^2 + \cdots + (n-1)^2\big) + 8\big(0 + 1 + \cdots + (n-1)\big) + n(-2) \\ &= 1 + 6\left[\frac{n(n-1)(2n-1)}{6}\right] + 8\left[\frac{n(n-1)}{2}\right] - 2n \\ &= 2n^3 + n^2 - 5n + 1. \end{aligned}$$

You should check that this function agrees with the table.

Solution for problem 11 on page 15.

(a) $x^3 - x^2 + 2x + 1$

(b) $-x^3 + 2x - 1$

(c) $x^3 - 1$

(d) $2x^4 + x^3 + 2x - 1$

(e) $5x^2 + 2x + 2$

(f) $1 + x + x^2 + x^3 + x^4 + x^5$

Notes for problem 13 on page 16. Using the method from this section, one polynomial function that agrees with this table is

$$\begin{aligned} f(n) &= \binom{n}{0} \cdot 3 + \binom{n}{1} \cdot -4 + \binom{n}{2} \cdot 5 + \binom{n}{3} \cdot 2 \\ &= \frac{1}{3}n^3 + \frac{3}{2}n^2 - \frac{35}{6}n + 3. \end{aligned}$$

(Is this the only *cubic* polynomial that fits the table? See problem 21 for the answer.)

Now, let's fill in the rest of the table using the "up and over" rule.

Input	Output	Δ	Δ^2	Δ^3	Δ^4
0	3	-4	5	2	0
1	-1	1	7	2	
2	0	8	9		
3	8	17			
4	25				
5					

Here, we also filled in the first entry in the Δ^4 column.

But note that the given information does *not* determine the output for $n = 5$. Thus, we can choose any value for $f(5)$. Setting $f(5) = 97$, for example, will give the following table:

Input	Output	Δ	Δ^2	Δ^3	Δ^4	Δ^5
0	3	-4	5	2	0	44
1	-1	1	7	2	44	
2	0	8	9	46		
3	8	17	55			
4	25	72				
5	**97**					

We now have *five* Δ columns corresponding with the *six* input values. This agrees with problem 12.

A polynomial function that fits this table is:

$$\begin{aligned} g(n) &= f(n) + \binom{n}{4} \cdot 0 + \binom{n}{5} \cdot 44 \\ &= \left[\frac{1}{3}n^3 + \frac{3}{2}n^2 - \frac{35}{6}n + 3\right] + \binom{n}{5} \cdot 44 \\ &= \frac{11}{30}n^5 - \frac{11}{3}n^4 + \frac{79}{6}n^3 - \frac{101}{6}n^2 + \frac{89}{30}n + 3. \end{aligned}$$

Changing $f(5)$ to another value will give us a different fifth degree polynomial.

Notes for problem 15 on page 16. The general proof uses mathematical induction on n and the recurrence formula

$$\binom{n}{k} = \binom{n-1}{k-1} + \binom{n-1}{k}.$$

Remember, focus on the *form* of the calculation rather than on its *value*.

If that becomes too messy, convince yourself that the repeated process of "up and over" will always yield the desired result.

Solution for problem 16 on page 20.

(a) $3x^3 - x^2 + 2x + 1$

(b) $x^3 + 2x - 1$

(c) $x^3 - 1$

(d) $2x^4 + x^3 + 2x - 1$

(e) $5x^2 + 2x + 2$

(f) $1 + x + x^2 + x^3 + x^4 + x^5$

Notes for problem 20 on page 22. Let

$$f(x) = a_0 + a_1 x + a_2 x^2 + \cdots + a_m x^m$$
$$g(x) = b_0 + b_1 x + b_2 x^2 + \cdots + b_m x^m$$
$$h(x) = f(x) - g(x)$$

Since h is a polynomial of degree at most m, it has at most m roots (by the Fundamental Theorem of Algebra). But f and g agree at $m + 1$ input, which imply that there are $m + 1$ values at which $h(x) = 0$. The only way to avoid this contradiction is to conclude that $h(x) \equiv 0$; that is, $h(x)$ is the zero polynomial. Thus, $f(x) = g(x)$ for all x.

Notes for problem 21 on page 22. To get a feel for this statement, let's look at Theorem 2 with $m = 3$. (This table is from problem 13.)

Input	Output	Δ	Δ^2	Δ^3
0	3	−4	5	2
1	−1	1	7	
2	0	8		
3	8			

Filling in the table gives us the above result. By Theorem 2,

$$f(x) = \sum_{k=0}^{3} a_k \binom{x}{k}$$

agrees with this table. Suppose g is another polynomial of degree less than or equal to 3 that also fits this table. Since f and g agree at four inputs ($n = 0, 1, 2, 3$), they must be identical by problem 20. Thus, f must be the unique polynomial of smallest degree that agrees with this table.

Notes for problem 26 on page 27. Let $f(x) = \binom{x}{k}$ and $g(x) = \binom{x-1}{k} + \binom{x-1}{k-1}$. Using the formula for $\binom{x}{k}$, we see that f and g have degree at most k.

But if n is any nonnegative integer, we have the identity

$$\binom{n}{k} = \binom{n-1}{k} + \binom{n-1}{k-1}$$

or $f(n) = g(n)$. Thus, f and g certainly agree at $k+1$ inputs. (In fact, they agree at infinitely many inputs!) Therefore, they must be identical polynomials.

Notes for problem 28 on page 28. Let's do some concrete examples to find a pattern. Using problem 27, we get

$$\Delta^1\left(\binom{x}{m}\right) = \binom{x}{m-1}$$

$$\Delta^2\left(\binom{x}{m}\right) = \Delta\left(\binom{x}{m-1}\right) = \binom{x}{m-2}$$

$$\Delta^3\left(\binom{x}{m}\right) = \Delta\left(\binom{x}{m-2}\right) = \binom{x}{m-3}$$

$$\vdots$$

$$\Delta^n\left(\binom{x}{m}\right) = \Delta\left(\binom{x}{m-n+1}\right) = \binom{x}{m-n} \qquad (*)$$

The formula $(*)$ holds for all $m \geq n$. We'll leave it to you to show that

$$\Delta^n\left(\binom{x}{m}\right) = 0 \quad \text{if } n > m.$$

Notes for problem 32 on page 28. Let n be the degree of f. Using problem 31 and the linearity of Δ (Theorem 4 on page 31), we can show that $\Delta^k(f)(x) = 0$ if $k > n$. Thus, the infinite sum

$$\sum_{k=0}^{\infty} \Delta^k(f)(0)\binom{x}{k}$$

is actually equal to the polynomial

$$g(x) = \sum_{k=0}^{n} \Delta^k(f)(0)\binom{x}{k}$$

whose degree is at most n. By Theorem 3, g agrees with f at $n+1$ inputs, namely when $x = 0, 1, \ldots, n$. Thus, they must be identical polynomials.

Notes for problem 42 on page 29. One of the important properties of the Δ operator is its *linearity*. That is, given polynomials f and g and a constant c, we have the following.

- $\Delta(f+g) = \Delta(f) + \Delta(g)$
- $\Delta(cf) = c\Delta(f)$

These will be discussed in the next section.

Notes for problem 43 on page 29. Let $f(x)$ be a polynomial of degree n and suppose $\Delta(f) = 0$. Assume for contradiction that $n \geq 2$ (we'll leave it up to you to show that n cannot be 1 either.) Using problem 30 and the linearity of Δ (Theorem 4), we can conclude that $\Delta(f)$ is a polynomial of degree $n - 1 \geq 1$. But this contradicts the fact that $\Delta(f) = 0$, and so f must be a constant function with $n = 0$.

We'll leave it to you to show that if f is a constant function, then $\Delta(f) = 0$.

Notes for problem 49 on page 34. If $\Delta f = \Delta g$, we have $\Delta f - \Delta g = 0$ or $\Delta(f-g) = 0$. Thus, by problem 43, $f-g$ is a constant and so $g(x) = f(x)+c$.

Notes for problem 58 on page 35. We proceed by induction on k. For the base case, suppose $k = 1$. Then the sum on the right-hand side is simply $f(x+1) - f(x)$, so the result is true if $k = 1$. Next, suppose it is true for $k - 1$; that is, suppose that

$$\Delta^{k-1}(f)(x) = \sum_{j=0}^{k-1}(-1)^j \binom{k-1}{j} f\big(x + (k-1) - j\big).$$

Then

$$\begin{aligned}
\Delta^k(f)(x) &= \Delta\big(\Delta^{k-1}(f)(x)\big) \\
&= \Delta\left[\sum_{j=0}^{k-1}(-1)^j \binom{k-1}{j} f\big(x + (k-1) - j\big)\right] \\
&= \Delta\left[\binom{k-1}{0} f\big(x + (k-1)\big) - \binom{k-1}{1} f\big(x + (k-2)\big)\right. \\
&\quad + \binom{k-1}{2} f\big(x + (k-3)\big) \\
&\quad \left. - \binom{k-1}{3} f\big(x + (k-4)\big) + \cdots\right] \quad \text{(now, use linearity)} \\
&= \binom{k-1}{0}\Big[f(x+k) - f\big(x + (k-1)\big)\Big] \\
&\quad - \binom{k-1}{1}\Big[f\big(x + (k-1)\big) - f\big(x + (k-2)\big)\Big] \\
&\quad + \binom{k-1}{2}\Big[f\big(x + (k-2)\big) - f\big(x + (k-3)\big)\Big] \\
&\quad - \binom{k-1}{3}\Big[f\big(x + (k-3)\big) - f\big(x + (k-4)\big)\Big] + \cdots \\
&= \binom{k-1}{0} f(x+k) - \left[\binom{k-1}{0} + \binom{k-1}{1}\right] f\big(x + (k-1)\big) \\
&\quad + \left[\binom{k-1}{1} + \binom{k-1}{2}\right] f\big(x + (k-2)\big)
\end{aligned}$$

$$-\left[\binom{k-1}{2}+\binom{k-1}{3}\right] f(x+(k-3))+\cdots$$
$$=\binom{k}{0} f(x+k)-\binom{k}{1} f(x+(k-1))$$
$$+\binom{k}{2} f(x+(k-2))-\binom{k}{3} f(x+(k-3))+\cdots$$
$$=\sum_{j=0}^{k}(-1)^j\binom{k}{j} f(x+k-j).$$

Notes for problem 63 on page 36. First, note that we need to define $a_{m,k}=1$ if $m=k=0$ and $a_{m,k}=0$ if $k>m$ or $k<0$ because $a_{m,k}=\Delta^k(f)(0)$ where $f(x)=x^m$ (see problem 59).

To show that $a_{m,k}=k(a_{m-1,k-1}+a_{m-1,k})$, try it in a case that's general in principle. Suppose, for example, $m=5$ and $k=4$. Then we have

$$a_{m-1,k-1}=a_{4,3}=\binom{3}{0}3^4-\binom{3}{1}2^4+\binom{3}{2}1^4,$$
$$a_{m-1,k}=a_{4,4}=4^4-\binom{4}{1}3^4+\binom{4}{2}2^4-\binom{4}{3}1^4.$$

Adding them together, we get

$$a_{m-1,k-1}+a_{m-1,k}=4^4-\left(\binom{4}{1}-\binom{3}{0}\right)3^4+\left(\binom{4}{2}-\binom{3}{1}\right)2^4$$
$$-\left(\binom{4}{3}-\binom{3}{2}\right)1^4$$
$$=4^4-\binom{3}{1}3^4+\binom{3}{2}2^4-\binom{3}{3}1^4.$$

Here, we used the identity $\binom{k}{j}-\binom{k-1}{j-1}=\binom{k-1}{j}$.

Now, multiply by $k=4$.

$$k(a_{m-1,k-1}+a_{m-1,k})=4^5-4\binom{3}{1}3^4+4\binom{3}{2}2^4-4\binom{3}{3}1^4$$
$$=4^5-\binom{4}{1}3^5+\binom{4}{2}2^5-\binom{4}{3}1^5$$
$$=a_{5,4}$$
$$=a_{m,k}.$$

The last simplification was done using the identity

Verify this on your own.

$$k\binom{k-1}{j}(k-j)^{m-1}=\binom{k}{j}(k-j)^m.$$

Notes for problem 64 on page 36. Let's see why this definition holds when $m = 4$. Note that

$$\binom{x}{4} = \frac{x(x-1)(x-2)(x-3)}{4!} = \frac{x(x-1)(x-2)}{3!} \cdot \frac{x-3}{4}$$
$$= \binom{x}{3}\frac{x-3}{4} = \frac{\binom{x}{3}x - \binom{x}{3}3}{4}.$$

We can conclude the following from the above: To find $b_{4,3}$ (the coefficient of x^3 in the expansion of $\binom{x}{4}$), take $b_{3,2}$ (the coefficient of x^2 in the expansion of $\binom{x}{3}$) minus 3 times $b_{3,3}$ (the coefficient of x^3 in the expansion of $\binom{x}{3}$), then divide by 4.

Therefore,

$$b_{4,3} = \frac{b_{3,2} - 3b_{3,3}}{4},$$

as desired.

Notes for problem 89 on page 44. Let

$$f(x) = \binom{x}{5} = \frac{x(x-1)(x-2)(x-3)(x-4)}{5!}.$$

Clearly, f vanishes at $x = 0, 1, 2, 3, 4$. Let g also be a polynomial that vanishes at $x = 0, 1, 2, 3, 4$. Then, by Theorem 10,

$$g - f = \lambda x(x-1)(x-2)(x-3)(x-4).$$

Thus,

$$g = f + \lambda x(x-1)(x-2)(x-3)(x-4)$$
$$= \left(\frac{1}{5!} + \lambda\right) x(x-1)(x-2)(x-3)(x-4)$$
$$= \left(\frac{1 + 5!\lambda}{5!}\right) x(x-1)(x-2)(x-3)(x-4)$$
$$= (1 + 5!\lambda)\frac{x(x-1)(x-2)(x-3)(x-4)}{5!}$$
$$= (1 + 5!\lambda)\binom{x}{5}$$

which is a multiple of $\binom{x}{5}$.

2

Form and Function: The Algebra of Polynomials

Introduction

In Chapter 1, our focus was on fitting polynomials to tables. In this chapter, we take polynomials themselves as the objects of study.

We're in a curious position with what we know about polynomials as a result of Chapter 1. We know that if we write a polynomial f as a combination of combinatorial polynomials, $\binom{x}{k}$, we have a simple interpretation of the numerical coefficients: Theorem 9 on page 37 of Chapter 1 says that if f is a polynomial of degree m, we can express f this way:

In case it's been a while since you've looked at Chapter 1, if f is a function, $\Delta(f)$ is the function defined by $x \mapsto f(x+1) - f(x)$. The numbers $\Delta^k(f)(0)$ come from the "0th" row of the difference table. See page 22 of Chapter 1 for a refresher for all this.

$$f(x) = \sum_{k=0}^{m} \Delta^k(f)(0)\binom{x}{k}.$$

So, if

$$f(x) = 4\binom{x}{3} - 13\binom{x}{2} + 5\binom{x}{1} - 6\binom{x}{0},$$

we know exactly where 4, -13, 5, and 6 come from:

Perhaps it's more interesting that you can go the other way, too: you can construct f from the 0th row of its difference table.

$$-6 = \Delta^0(f)(0) = f(0)$$
$$5 = \Delta^1(f)(0)$$
$$-13 = \Delta^2(f)(0)$$
$$4 = \Delta^3(f)(0)$$

and the difference table for f starts out:

n	$f(n)$	$\Delta(f)(n)$	$\Delta^2(f)(n)$	$\Delta^3(f)(n)$	$\Delta^4(f)(n)$	$\Delta^5(f)(n)$	$\Delta^6(f)(n)$	$\cdots$
0	-6	5	-13	4	0	0	0	$\cdots$
1	$\vdots$	$\vdots$	$\vdots$	$\vdots$	$\vdots$	$\vdots$	$\vdots$	

This is the same f as the one above, just written in a different form.

But what if you write a polynomial in the "normal" form? Say

$$f(x) = \frac{2}{3}x^3 - \frac{17}{2}x^2 + \frac{77}{6}x - 6.$$

What interpretation can you give to the numbers

$$\frac{2}{3}, \quad \frac{17}{2}, \quad \frac{77}{2}, \quad \text{and } -6?$$

In this chapter, we'll answer this question in two ways: More precisely,

- We'll relate the coefficients of a polynomial to the values of the function (and the values of its derivatives).
- We'll see how the derivative arises in a completely algebraic way, with no recourse to calculus.
- We'll also see how the coefficients are related to the zeros of the function—generalizing the "sum and product of the roots" topics from algebra.

To be precise, functions have zeros and equations have roots, but we'll use the terms "zero" and "root" interchangeably.

More generally, we'll look into how changing the form of a polynomial expression gives you insight into the underlying function.

2.1 Polynomials

Let's be a little more precise about what we mean by the word "polynomial." The word is used in two closely related ways in school mathematics:

- **Polynomial *Functions*:** When you think about the "letter" in a polynomial as a *variable*, you are thinking of the polynomial as a machine that takes inputs to outputs. It generates a table, has a graph, and has all the other attributes of real-valued functions of a real variable. This is the point of view underneath graphing and tabulation (spreadsheet) technology. Using this point of view, two polynomial functions are equal if they represent the *same function*: They produce the same output for any input—they always tabulate the same and they have the same graph.
- **Polynomial *Forms*:** When you think of a polynomial as a formal expression, the "letter" is an *indeterminate*. This is the "algebra 2" perspective that you use when you factor, add, multiply, combine like terms, and so on. It's the point of view underneath CAS technology. Looked at as forms, two polynomials are equal if, when written in "standard" form, they have the same degree and the same coefficients. Of course, you may have to use some algebra to get them into the same standard form.

The mathematician Herman Weyl put it this way:

This is from an essay that Weyl wrote in 1931, entitled "Topology and Abstract Algebra as Two Roads of Mathematical Comprehension."

"The system of real numbers is like a Janus head with two oppositely directed faces. In one respect it is the domain of $+$ and $\times$ and their inverses, and in another it is a continuous manifold, and the two are continuously related. One is the algebraic and the other is the topological face of numbers. The idea that the argument x is a variable that traverses continuously its values is foreign to algebra; it is just an indeterminate, an empty symbol that binds the coefficients of the polynomial into a uniform expression that makes it easier to remember the rules for addition and multiplication. 0 is the polynomial all of whose coefficients are 0 (not the polynomial that takes on the value 0 for all values of the variable x)."

For our purposes, form and function represent two different ways to *think about* polynomials.

Let's adopt the convention that "normal" means that you write the polynomial as a sum of monomials whose powers go from largest to smallest, like $5x^4 - 3x^3 + 3x^2 - 2$. Your CAS may use a different convention, but it should be easy to translate its normal form to this one.

- When you think of forms, you are thinking about algebraic calculations and transformations with polynomials, forgetting about the meaning of the variable, and establishing *identities*. An identity is established by showing that each side can be transformed into the same "normal form" using the rules of algebra.
- When you think about functions, you worry about tables and graphs, notions like continuity, slopes of tangent lines (rates of change), the possible numerical *values* you can obtain, and, what's especially important in this chapter, the properties of the function that you can figure out by substituting numbers and expressions for the variable.

Form and function come together in the *theory of equations*. Sometimes, when you change the form of a polynomial expression, you gain insight into properties of the associated function. Writing $f(x) = x^2 - 1$ as $f(x) = (x+1)(x-1)$ tells you what inputs produce 0 as an output.

It's certainly true that polynomials that are equal as forms are equal as functions. Indeed, if you have a polynomial identity, you have infinitely many numerical identities, because you can substitute numbers for the variable. So, since $x^2 - 1 = (x-1)(x+1)$ *as polynomials*, we know that

Let's call this the "form implies function" property of polynomials.

- $5^2 - 1 = 4 \cdot 6$,
- $8^2 - 1 = 7 \cdot 9$,
- $900^2 - 1 = 899 \cdot 901$,
- $3 - 1 = (\sqrt{3} - 1)(\sqrt{3} + 1)$,

and so on. In fact, you can substitute *expressions* (other polynomials, for example) for x and obtain other identities:

- $(a+2)^2 - 1 = (a+1)(a+3)$,
- $(3x^2 - 4x + 4)^2 - 1 = (3x^2 - 4x + 3)(3x^2 - 4x + 5)$,
- $(x^8 - 1)^2 - 1 = (x^8 - 2)x^8$.

Some people describe this "true under any substitution" property as the "universal property" of identities.

Because we are working over $\mathbb{R}$, the converse is also true: If two polynomials define the same function, then they are equal as polynomials, and we'll prove that in section 2.2. But this isn't the case over every number system.

"Function implies form" (if we are working over $\mathbb{R}$).

The distinctions between polynomial forms and functions tend to be ignored in school mathematics. Getting the distinctions straight has made it easier to emphasize the right things at the right times with students, even if they are never asked to "learn" the precise distinctions for a test. For example, students often wonder about the importance of the quadratic formula, given the existence of calculators that will approximate roots of equations to any reasonable accuracy.

If you care about the *value* of your answer, numerical approximation may be good enough. If you care about its *form*, approximations will not do the trick.

Well, the quadratic formula tells us about the *algebraic* character of the roots. The numerical routines in the calculator give us information about the *values* of the roots. When looking for the numbers that make $x^2 - x + 1$ equal to 0, 10-place accuracy is probably all you'd ever need. But if you want to use the roots of this equation to generate and prove properties of Fibonacci numbers (see problems 28–30), the *algebraic* properties of the roots are what you care about.

Example: Here are two algebraic expressions:

$$f(x) = \frac{x^3 - 3x^2 - 4x + 12}{x - 2} \qquad \text{and} \qquad g(x) = x^2 - x - 6.$$

Are they equal? It depends on your frame of mind:

In abstract algebra, people say that "$f = g$ in the field of rational expressions over $\mathbb{R}$."

- If you look at them as formal expressions, they *are* equal because "equal" in this context means you can transform one into the other using legal moves:

$$\begin{aligned} \frac{x^3 - 3x^2 - 4x + 12}{x - 2} &= \frac{(x-2)(x+2)(x-3)}{x-2} \\ &= (x+2)(x-3) \\ &= x^2 - x - 6. \end{aligned}$$

People sometimes say that "$f = g$ except at 2." The graph of f has a hole above 2, but the graph of g is a continuous curve.

- If you look at them as functions, they are *not* equal because they have different domains: $g(2) = -4$, but $f(2)$ is undefined.

Ways to think about it

In abstract algebra, what we are calling a "system" is known as a *ring*.

Let's look at this a little more abstractly. We have two systems, polynomial forms in one variable with real coefficients (usually denoted by $\mathbb{R}[x]$), and polynomial *functions* on $\mathbb{R}$ that come from these polynomial forms (let's denote the system of these functions by $\mathbb{R}\langle x \rangle$).

These are *systems* rather than *sets*, because each of them allows for addition and multiplication: You can add and multiply polynomial forms and you can add and multiply polynomial functions:

- The way you add and multiply forms is, well, *formally*: You add expressions term by term, and you multiply expressions using the general distributive law, collecting like terms at the end.
- The way you add and multiply functions has nothing to do with the form used to represent the functions; the operations are carried out *functionally*—that is, by telling what the sum and product do to an input. And what they do is defined by two equations:

$$(f+g)(x) = f(x) + g(x), \quad \text{and} \quad (fg)(x) = f(x) \cdot g(x).$$

Technically, two functions are equal if they have the same domain and produce the same output for each input in that domain. Since polynomial functions all have the same domain (all of $\mathbb{R}$), we don't need to mention it in the definition of addition and multiplication.

(continued)

These are subtle equations. In the first one, for example, the "+" means different things on both sides of the equation—on the left, it's the sum of two *functions*, and on the right, it's the sum of two *outputs* (numbers in our case). The equation defines addition of functions in terms of addition of outputs.

Anyway, back to the abstractions: What we have is a correspondence between $\mathbb{R}[x]$ and $\mathbb{R}\langle x\rangle$, a function of sorts, that associates each polynomial form with a polynomial function. If we wanted to be fancy, we could name this correspondence, say, Λ and write

$$\Lambda : \mathbb{R}[x] \to \mathbb{R}\langle x\rangle$$

where Λ is defined by

$$\Lambda(f) = (x \mapsto f(x)).$$

This says that Λ of a polynomial form f is the function that assigns a number to what you get when you substitute the number for x in the polynomial.

Most of what has been discussed in this section can be stated in terms of Λ:

1. Λ is *well defined*. That means that a polynomial form can give rise to only one polynomial function ("form implies function").
2. Λ is *one-to-one*. That means that two different polynomial forms can't give rise to the same polynomial function ("function implies form").
3. Λ *preserves the structure* of $\mathbb{R}[x]$ and $\mathbb{R}\langle x\rangle$. This means that if you add two forms and look at the function they define, you get the same function as adding the functions defined by each form ("Λ of a sum" is the same as "the sum of the Λs"). Mathematicians like to illustrate this with what they call a "commutative diagram"—we'll spare you the details. And the same holds for multiplication.

More details about the Λ correspondence can be found in [2].

Facts and Notation

- For the rest of the book, when we say "polynomial," we'll usually mean *polynomial form*.
- When we want to think about a polynomial as a *function*, we'll usually say that—"let f be a polynomial function," for example.
- When we say a property of polynomials is *algebraic* or is "part of algebra" we mean that it can be derived using formal calculations of the kind you find on a CAS, thinking of polynomial forms.
- When we say a property of polynomials is *analytic* or is "part of analysis" we usually mean that it depends on thinking of polynomials as functions.

A sentence like "let f be a polynomial" usually means we are thinking of forms. If we write "$f(x)$," it's usually a context clue that we're thinking of functions. None of these conventions is hard and fast.

We'll be a little sloppy about this "one variable" restriction, because sometimes we'll want to replace "x" by "$x+a$" or even "$x+y$." In Chapter 3, we'll allow complex coefficients.

Unless, of course, $x+2$ is a factor of $5(x-4)^4+7x^2-1$. Is it?

Some of the problems are just fun. As you do them, ask yourself "am I thinking *form* or *function* or *both*?"

- Both flavors of polynomials will mean "polynomials in *one* variable with *real* coefficients." This includes things like

$$3x^2-\frac{5}{2}x+1, \quad \sqrt{3}y+y^5, \quad \text{and} \quad \frac{5(x-4)^4+7x^2-1}{\pi},$$

but not things like

$$3y^2-\frac{5}{2}x+1, \quad \sqrt{3y}+y^5, \quad \text{and} \quad \frac{5(x-4)^4+7x^2-1}{x+2}.$$

The next problem set provides some examples for the two ways to think about polynomials. Not all the problems are needed in what follows, but some of them are previews of ideas that will be fully developed later in the chapter. We'll refer back to these as we go on, so if you want to skip around in this set, you can do the appropriate problems as they are cited later.

Problems

1. Express each result in normal form:

$f(x)$	$f(2)$	$f(a)$	$f(a+2)$
$6x^2-x-2$			
x^2+x-1			
x^2+x+1			

2. Write each answer as a polynomial in x with coefficients that are polynomials in a:

$f(x)$	$f(x+a)$	$f(x)+f(a)$	$f(x-a)$	$f(x)-f(a)$	$f(ax)$	$f(a)f(x)$
$6x^2-x-2$						
x^2+x-1						
x^2+x+1						

3. For each function f, find

$f(x)$	The numbers α such that $f(\alpha)=0$	The numbers α such that $f(\alpha)=3$	The smallest and largest values of $f(x)$
$6x^2-x-2$			
x^2+x-1			
x^2+x+1			

4. Suppose

$$f(x,y)=(x+y+1)^5-x^5-y^5-1.$$

(a) Show that

$$f(x, y) = 5(y+1)(x+1)(x+y)(x^2+xy+y^2+x+y+1).$$

(b) Write f as a polynomial in x with coefficients that are polynomials in y.

(c) Write f as a polynomial in y with coefficients that are polynomials in x.

(d) For what numbers a is $f(x, a)$ identically 0?

(e) For what numbers a is $f(a, x)$ identically 0?

5. Suppose $f(x, y)$ and $g(x, y)$ are two polynomials in x and y, and we have the polynomial identity

$$f(x, y) = g(x, y).$$

Suppose also that $a_1, a_2, a_3, \ldots$ is a sequence of numbers. If, in f and g, we replace each power of y, y^k, with a_k, we get two polynomials in one variable, x. Show that these, too, are identically equal.

6. Let's let $\deg(f)$ denote the degree of the polynomial f (assume f is a polynomial in one variable x). If f and g are polynomials, find a relationship among

Try some concrete examples to find a pattern.

(a) $\deg(fg)$, $\deg(f)$, and $\deg(g)$.

(b) $\deg(2g)$ and $\deg(g)$.

(c) $\deg(f+g)$, $\deg(f)$, and $\deg(g)$.

7. What are the polynomials of degree 0?

8. What about the number 0? The convention is to call the degree of 0 "$-\infty$." Explain why this is a good idea.

Hint: Look at your solution to problem 6a.

9. Show that a polynomial of degree n can have at most n linear factors (that is, factors of the form $ax+b$).

10. Let f be any polynomial function. What is the domain of f?

11. If f and g are any functions, $f \circ g$ is the function defined by

$$(f \circ g)(x) = f\big(g(x)\big).$$

(a) Find $h \circ k$ if $h(x) = x^3 - 1$ and $k(x) = x^2 + 1$.

(b) Show that, if f and g are polynomial functions, so is $f \circ g$.

How is the polynomial $f \circ g$ obtained from the polynomials f and g?

(c) Show that, in general, $f \circ g \neq g \circ f$.

(d) Find two polynomials f and g so that $f \circ g = g \circ f$.

(e) True or false? $\deg(f \circ g) = \deg(g \circ f)$.

(f) If $r(x) = 2x^2 - 1$ and $s(x) = 5x - 20x^3 + 16x^5$, show that $r \circ s = s \circ r$.

In Chapter 3, we'll see how to construct "commuting" polynomials like r and s. It's connected to trigonometry.

Here Δ is the usual difference operator, a function in its own right. This problem shows that, if you're careful, you can do "algebra" with powers of Δ, thinking of $+$ and $\circ$ as the $+$ and $\times$ of first-year algebra.

12. Show that

$$(\Delta + \Delta^2) \circ (\Delta^3 + \Delta^4) = \Delta^4 + 2\Delta^5 + \Delta^6.$$

We'll use the shorthand "the graph of f" to mean the graph of the equation $y = f(x)$: the set of all points in the Cartesian plane of the form $(x, f(x))$, as x runs over the real numbers.

13. If f and g are polynomial functions,

(a) How is the the graph of $y = f(x) + g(x)$ related to the graphs of f and g?

(b) How is the the graph of $y = 2f(x)$ related to the graph of f?

(c) How is the the graph of $y = f(x) \cdot g(x)$ related to the graphs of f and g?

(d) How is the the graph of $y = f \circ g(x)$ related to the graphs of f and g?

Assume the graph doesn't do anything strange outside the shown window.

14. Name at least two polynomial functions that could have each picture as its graph.

(a)

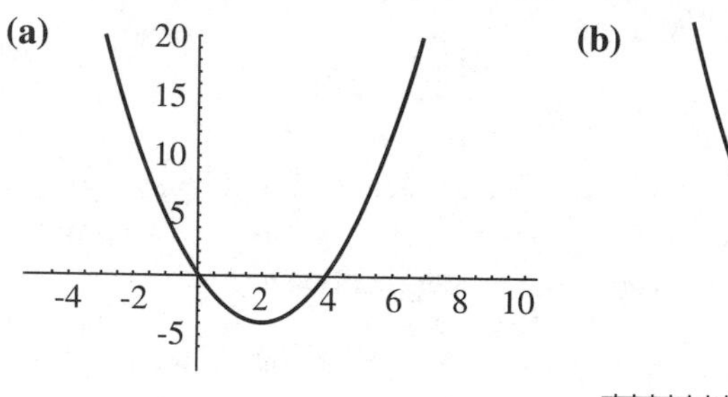

(b)

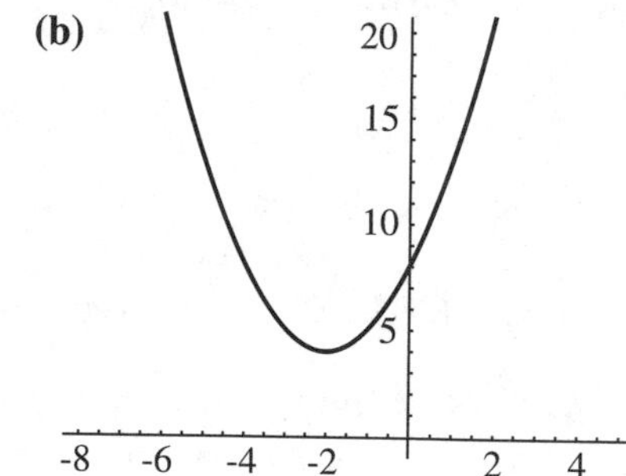

(c)

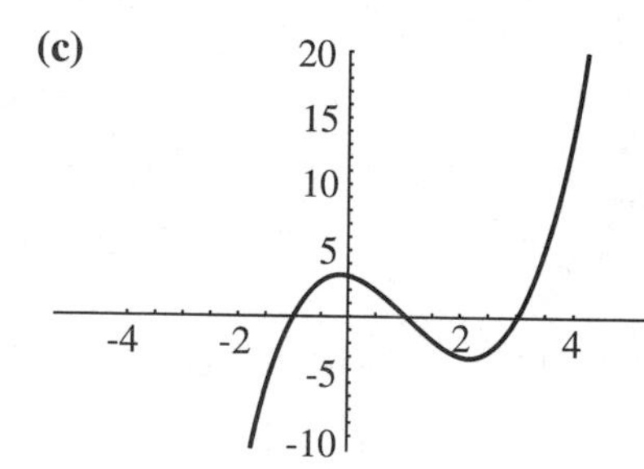

(d)

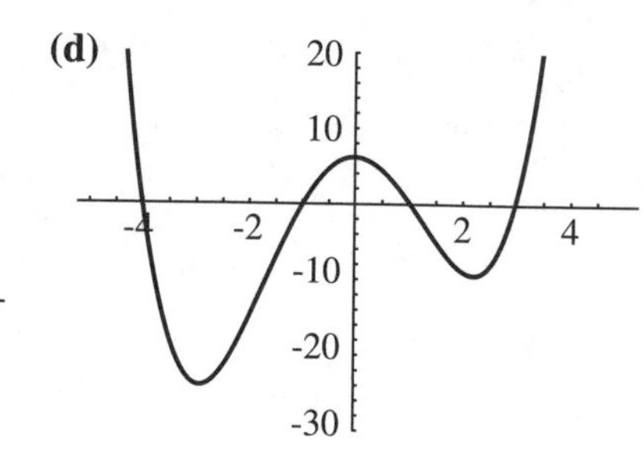

15. Show that the square of an odd number is always one more than a multiple of 4.

If you project the graph of f onto the y-axis, you'll cover the axis completely.

16. Suppose f is a polynomial function with the property that every real number is an output of f. What can you say about $\deg(f)$? Proof?

17. Show that a polynomial (with real coefficients) of odd degree has a real zero.

i.e., one-to-one

18. What degrees are possible for polynomials f that have the property that whenever $f(a) = f(b)$, $a = b$?

19. If n is a positive integer, what is the normal form for this product?

$$(x - 1)(1 + x + x^2 + \cdots + x^{n-1}).$$

Hint: Use problem 19 with $x = 2$.

20. If you earn one cent today and you double your daily rate every day, how many days will it take to earn more than $1000?

21. Suppose a number r, written in base 10, is a string of 100 "8"s. What is the highest power of 10 that is a factor of $9r + 8$?

22. Let $f(x) = 1 + x + x^2 + x^3 + x^4$. Show, without expanding, that all the coefficients of $f(x + 1)$ except the leading one are divisible by 5.

Hint: Use problem 19 and the Binomial theorem.

23. If n is a positive integer, simplify

$$(x + 1)(1 - x + x^2 - x^3 \pm \cdots \pm x^{n-1}).$$

Remember, "simplify" means "put in normal form." In this case, just expand the product and write the the terms in descending order.

24. If n is a positive integer, simplify

$$(x + 1)(x^2 + 1)(x^4 + 1)(x^8 + 1) \cdots (x^{2^n} + 1).$$

25. Factor over $\mathbb{R}$:

$$x^4 + x^2 + 1.$$

26. Prove the *quadratic formula*: The roots to $ax^2 + bx + c = 0$ are

$$\frac{-b + \sqrt{b^2 - 4ac}}{2a} \quad \text{and} \quad \frac{-b - \sqrt{b^2 - 4ac}}{2a}.$$

27. There's something special about a 72° angle: There's only one isosceles triangle (up to similarity) whose base angle is twice the vertex angle. It is the "72-72-36 triangle."

For geometry teachers: One consequence of this problem is that the side of a regular decagon inscribed in a circle can be constructed with straight-edge and compass. So the regular decagon, 20-gon, and 10n-gon are all constructible with these tools. So is the pentagon (connect every other vertex of the decagon). How long is the side of the pentagon if the circle's radius is taken to be 1?

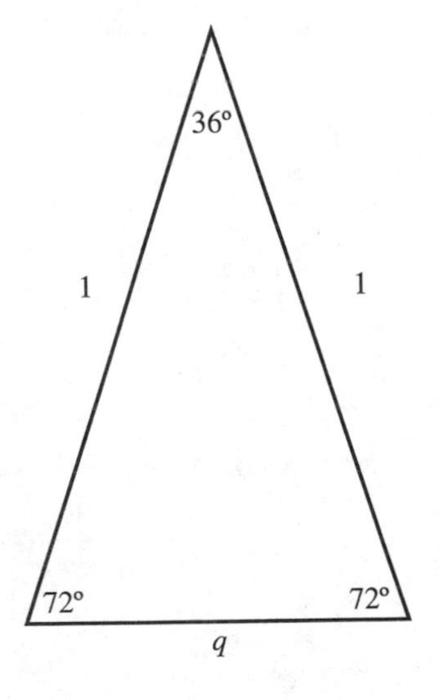

Suppose the equal sides of the triangle have length 1, and let q stand for the length of the base.

(a) Bisect one of the base angles of the triangle.

(b) Show that the small triangle is similar to the whole triangle.

(c) Use (b) to show that

$$\frac{1}{q} = \frac{q}{1 - q}.$$

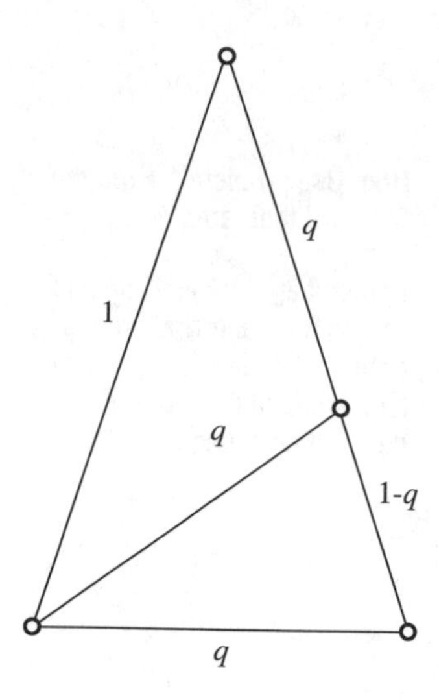

(d) Show that $q = 2\cos 72°$.

(e) Solve for q and for $\cos 72°$.

28. The Fibonacci numbers are the outputs of the function defined by the rule

$$f(n) = \begin{cases} 1 & \text{if } n = 0 \text{ or } 1, \\ f(n-1) + f(n-2) & \text{if } n > 1. \end{cases}$$

These numbers have many wonderful properties (see [11] for some of them). The sequence

$$\{f(0), f(1), f(2), f(3), f(4), f(5), f(6), \ldots\} = \{1, 1, 2, 3, 5, 8, 13, \ldots\}$$

is neither geometric (constant ratio between terms) nor arithmetic (constant difference between terms). But it still exhibits some regularity. Write out a few terms of the sequence of ratios

$$\left\{ \frac{f(1)}{f(0)}, \frac{f(2)}{f(1)}, \frac{f(3)}{f(2)}, \frac{f(4)}{f(3)}, \frac{f(5)}{f(4)} \cdots \right\}.$$

If you write these fractions as decimals, they seem to converge. What's a good approximation to the limit? What's the exact value of the limit?

Hint: $\beta = \frac{1-\sqrt{5}}{2}$ is also a root of this polynomial.

29. Find a polynomial with integer coefficients for which

$$\alpha = \frac{1+\sqrt{5}}{2}$$

is a root.

Tabulate it between 1 and 20, for example.

Hint: If $\alpha = \frac{1+\sqrt{5}}{2}$, $\alpha^2 = \alpha + 1$.

30. Experiment with this function:

$$g(n) = \frac{1}{\sqrt{5}}\left(\left(\frac{1+\sqrt{5}}{2}\right)^n - \left(\frac{1-\sqrt{5}}{2}\right)^n\right).$$

Show that g produces the Fibonacci numbers as outputs.

These are special cases of the "sum of roots" and the "product of roots" formulas.

31. Show that, if α and β are roots of the quadratic equation $x^2 + bx + c = 0$, then

(a) $\alpha + \beta = -b$,

(b) $\alpha\beta = c$.

32. Show that

(a) $(x - r)(x - s) = x^2 - (r + s)x + rs$,

(b) $(x - r)(x - s)(x - t) = x^3 - (r + s + t)x^2 + (rs + rt + st)x - rst$.

33. Generalize the result of problem 32 to an identity that expands

$$(x - \alpha_1)(x - \alpha_2)(x - \alpha_3) \cdots (x - \alpha_n).$$

This identity will be useful in section 2.5.

34. Show that

$$(r + s)^3 - 3rs(r + s) = r^3 + s^3.$$

35. Consider the polynomials:

$$\{x^2 - 1, x^3 - 1, x^4 - 1, x^5 - 1, \ldots, x^n - 1, \ldots\}.$$

(a) Investigate how these polynomials factor over the integers.

(b) Make a table:

n	number of factors of $x^n - 1$
1	
2	
3	
4	
5	
6	
$\vdots$	
20	

(c) For what values of n does it seem that $x^n - 1$ factors into *exactly* two factors?

(d) Find a function that agrees with the above table.

Beware: the simplest function may not be polynomial; you might want to describe its rule in words.

2.2 The Basic Theorems

In Chapter 1, we stated several facts from algebra that may or may not be familiar to you. The purpose of this section is to sketch proofs of these facts. The facts themselves will be useful throughout the rest of this book. If you already know proofs, see if your proofs are the same as ours.

Theorem 1. (Long division) *If f and g are polynomials, with $g \neq 0$, there exist polynomials q and r such that the degree of r is strictly less than the degree of g and*

$$f = gq + r.$$

We call this "long division" because the proof of the theorem is based on the usual algorithm for dividing one polynomial into another. Consider this example:

$$\begin{array}{r|l} & 3x + \frac{5}{2} \\ 2x^2 - x - 1 & 6x^3 + 2x^2 + x - 1 \\ & 6x^3 - 3x^2 - 3x \\ \hline & 5x^2 + 4x - 1 \\ & 5x^2 - \frac{5}{2}x - \frac{5}{2} \\ \hline & \frac{13}{2}x + \frac{3}{2} \end{array}$$

You first find a trial quotient, namely $3x$ ("how many times does $2x^2$ go into $6x^3$?"), multiply the divisor by the quotient, subtract (killing the leading term of the "dividend"), and repeat, until the degree of the divisor is larger than the degree of the dividend. Now, to make this all precise might seem like more trouble than it's worth, but, in case you'd like to see the gory details, here's a "formal" proof:

The case $m < n$ is like a fact that confuses many students in arithmetic: The remainder when 37 is divided by 72 is 37 itself (and the quotient is 0).

Proof Suppose

$$f(x) = a_n x^n + a_{n-1}x^{n-1} + \cdots + a_0$$

and

$$g(x) = b_m x^m + b_{m-1}x^{m-1} + \cdots + b_0.$$

We'll use induction on m, the degree of g. If $m < n$, the theorem is trivially true because

$$g = 0 \cdot f + g \quad \text{and} \quad \deg(g) < \deg(f).$$

So, suppose $m \geq n$ and the theorem is true for every polynomial h of degree less than m. Then let

$$h(x) = g(x) - \frac{b_m}{a_n} f(x) x^{m-n}.$$

Try to see how this really is just a formalization of what you do when you do long division.

If you write this all out, you see that this construction kills off the leading term in g so that $\deg(h) < m$. By the induction hypothesis, there are polynomials $s(x)$ and $r(x)$ so that

$$h(x) = s(x)f(x) + r(x) \quad \text{and} \quad \deg(r) < \deg(f).$$

The fact that you can do division with remainder for polynomials (that's the basic conclusion in Theorem 1) forms the basis for the structural similarity between polynomials and integers. It's the reason you can talk about factorizations, greatest common divisors, and the rest of arithmetic in both systems.

But then

$$g(x) - \frac{b_m}{a_n} f(x) x^{m-n} = s(x)f(x) + r(x)$$

so

$$g(x) = \left(\frac{b_m}{a_n} x^{m-n} + s(x)\right) f(x) + r(x).$$

If we let

$$q(x) = \left(\frac{b_m}{a_n} x^{m-n} + s(x)\right)$$

Note that the $r(x)$ is the same whether the dividend is h or g.

we have

$$g(x) = q(x)f(x) + r(x) \quad \text{and} \quad \deg(r) < \deg(f)$$

as advertised. ■

Theorem 2. (The remainder theorem) *If f is a polynomial, then the remainder when $f(x)$ is divided by $x - a$ is $f(a)$.*

We're assuming here that a is a number.

Proof Apply Theorem 1 to f and $x - a$. It says that

$$f(x) = q(x)(x - a) + r(x) \quad \text{and} \quad \deg(r) < \deg(x - a) = 1.$$

Well, the only polynomials of degree < 1 are the constants, so $r(x)$ is a *number*, call it r. So,

$$f(x) = q(x)(x - a) + r.$$

Now substitute a for x; you get

$$f(a) = q(a)(a - a) + r = q(a) \cdot 0 + r.$$

So, $r = f(a)$. ■

Theorems 1 and 2 are the basic results we need. Several useful corollaries follow:

Corollary 1. (The factor theorem) *If f is a polynomial, $(x - a)$ is a factor of $f(x)$ if and only if $f(a) = 0$.*

The factor theorem is the classic example of form and function coming together: "$(x - a)$ is a factor of $f(x)$" is a statement about polynomial algebra, and "$f(a) = 0$" talks about f as a function.

The proof is an exercise (problem 36).

Corollary 2. (The number of roots theorem) *A polynomial f of degree n can have at most n zeros.*

Proof Every root produces a linear factor, and every linear factor increases the degree of the product of factors by 1. You can make this precise using an argument similar to that in the proof of Theorem 10 on page 43 of Chapter 1. ■

Note that this proof breaks down if $f = 0$ (that is, if $\deg(f) = -\infty$), so a variation on this corollary is

Corollary 2. (Alternate version) *A polynomial f with more roots than its degree is identically 0.*

Corollary 3. (The function implies form theorem) *If two polynomials of degree at most n agree for more than n inputs, they are exactly the same polynomial.*

This is the converse of the "form implies function" property described on page 55.

Proof If f and g agree at $m > n$ numbers $\alpha_1, \ldots, \alpha_m$, then $f - g$, a polynomial of degree at most n, vanishes at m numbers, so it is identically 0. If $f - g = 0$, $f = g$. ■

Corollary 4. *A polynomial of degree n is completely determined by any $n + 1$ of its values.*

Proof Two polynomials of degree n that agree at these $n+1$ values have to be identical by Corollary 3.

Example: Suppose we are looking for a simple (polynomial) formula that lets you calculate the sum of the first n squares. That is, we want a polynomial $f(x)$ so that

$$f(n) = 0^2 + 1^2 + \cdots + (n-1)^2.$$

The corresponding polynomial for the sum of the *first* powers is $\frac{x(x-1)}{2} = \binom{x}{2}$ because $0+1+2+\cdots+n-1 = \frac{n(n-1)}{2}$.

Imagine we had such a polynomial. Could we determine its degree? Well (in the spirit of Chapter 1), on positive integers, f would tabulate like this:

n	$f(n) = 0^2 + 1^2 + \cdots (n-1)^2$	$\Delta(f)(n)$
1	0	1
2	1	4
3	5	9
4	14	16
5	30	25
6	55	36
7	91	

Remember the hockey stick property?

n	$f(n)$	$\Delta(f)(n)$
1	0	1
2	1	4
3	5	9
4	14	16
5	30	25
6	55	36
7	91	

Look at the Δ column. It looks like perfect squares. In fact, if n is any positive integer,

$$\begin{aligned}\Delta(f)(n) &= f(n+1) - f(n)\\ &= \left(0^2 + 1^2 + \cdots n^2\right) - \left(0^2 + 1^2 + \cdots + (n-1)^2\right)\\ &= n^2.\end{aligned}$$

And, if $\Delta(f)(x) = x^2$ and $f(1) = 0$, the hockey stick property of Chapter 1 implies that

$$f(n) = 0 + 1 + 4 + 9 + \cdots + (n-1)^2$$

and that's what we want. So, we're on the hunt for a polynomial f so that $f(1) = 0$ and $\Delta(f)(x) = x^2$. But, by problem 73 on page 37 of Chapter 1, this implies that f must have degree 3. We know that a cubic is determined by four values (Corollary 4), so fit a cubic to this table:

"Fit a cubic" using the methods of Chapter 1, for example.

Input	Output
1	0
2	1
3	5
4	14

If you stumbled on an f so that $\Delta(f)(x) = x^2$ but $f(1) \neq 0$, you could subtract the number $f(1)$ from f and not ruin the fact that $\Delta(f)(x) = x^2$ (see problem 48 on page 34 of Chapter 1).

We get

$$f(x) = \frac{1}{3}x^3 - \frac{1}{2}x^2 + \frac{1}{6}x = \frac{x(x-1)(2x-1)}{6}.$$

You can check that $\Delta(f)(x) = x^2$. Since $f(1) = 0$, we have the nice formula

$$0^2 + 1^2 + \cdots + (n-1)^2 = \frac{n(n-1)(2n-1)}{6}.$$

Ways to think about it

We're being a little cavalier about the way we use words like "root" and "number." Even if we restrict the coefficients of our polynomials to be real numbers, the zeros of the corresponding functions may well turn out to be complex numbers. We'll deal explicitly with complex numbers in the next chapter, but for now, we have to allow them to lurk behind the scenes. For example, in problem 31 on page 62, the sum and product of the roots are what the problem claims they are (nice real numbers), even if the roots are complex.

In fact, in section 2.6, we'll make implicit use of a sort of converse to the "number of roots theorem" (Corollary 2), the celebrated *fundamental theorem of algebra* that says that every polynomial of degree n with real (or even complex) coefficients factors into exactly n linear factors in the complex numbers. The FTA implies that every polynomial equation of degree n has n roots in the complex numbers (if we count multiple roots as distinct). So, we're safe in assuming that a polynomial *has* roots, and we know how many it has, as long as we assume the FTA.

The FTA is not so easy to prove. All proofs of the theorem need a so-called "analytic step:" a result that's part of analysis and not strictly algebra. For example, the fact that a polynomial of odd degree has a real root (problem 17 on page 60) is such an analytic step, one that is often used in proofs of the FTA.

Problems

36. Prove Corollary 1.

37. Prove that a polynomial with more roots than its degree is identically 0 (the "alternate version" of Corollary 2).

38. Prove that two polynomials that agree infinitely often are identical.

39. Find the remainders when each polynomial is divided by $x - 3$:

Hint: Is long division really necessary?

(a) $x^3 - 1$ **(b)** $x - 1$
(c) $x^2 + x + 1$ **(d)** $x^4 + x^2 + 1$
(e) $(x^4 + x^2 + 1)(x - 1)$ **(f)** $x^4 + x^3 + x^2$

40. Suppose f and g are polynomials and that $f(2) = 5$ and $g(2) = -4$. Find the remainder when each polynomial is divided by $x - 2$:

(a) $3f + g$ **(b)** fg^2
(c) $(x^2 + x + 1)f(x) + g(x)$

Is the converse of problem 41 true? See [13] for more on this theme.

41. Suppose $f(S, C)$ is a polynomial in two variables that vanishes for all pairs of numbers (S, C) such that $S^2 + C^2 = 1$. Show that there is a polynomial $g(S, C)$ so that

$$f(S, C) = (S^2 + C^2 - 1)g(S, C).$$

Problems 42–44 outline a different proof of the remainder theorem.

This generalizes problem 19 on page 60.

42. Establish the identity

$$x^n - a^n = (x - a)(x^{n-1} + ax^{n-2} + a^2x^{n-3} + a^3x^{n-4} + \cdots + a^{n-2}x + a^{n-1}).$$

43. Suppose

$$f(x) = a_nx^n + a_{n-1}x^{n-1} + \cdots + a_1x + a_0.$$

Show that

$$f(x) - f(a) = (x - a)g(x)$$

for some polynomial g (in x and a), and write down an explicit expression for g.

44. Use problem 43 to prove the remainder theorem.

45. Find a formula for the sum of the first n cubes, starting with 0:

$$0^3 + 1^3 + 2^3 + \cdots + (n-1)^3.$$

2.3 Coefficients and Values

We're writing our generic polynomial in "reverse" normal form, just because it makes some formatting easier later in this section.

Let's take up the question we asked on page 54: What if we write a polynomial, not as a linear combination of the $\binom{x}{k}$, but as a linear combination of the "standard basis" (the x^k)? Suppose

$$f(x) = c_0 + c_1x + c_2x^2 + \cdots + c_mx^m.$$

What can you say about the c_k? Are they related to anything like the "$\Delta^k(f)(0)$"? Well, yes, and the exact analogy might be a bit of a surprise. Here's the story.

One of the things you need to do when you calculate $\Delta(f)$ is to figure out $f(x + 1)$. The question here is more general: what is a formula for $f(x + a)$?

When you teach about polynomial functions, you probably ask students, given a formula for $f(x)$, to find a formula for $f(-x)$, $f(x + 2)$, or maybe even $f(x + a)$.

Example: Suppose $f(x) = 1 - 4x + 5x^2 + 3x^4$. Then

$$\begin{aligned} f(x + a) &= 1 - 4(x + a) + 5(x + a)^2 + 3(x + a)^4 \\ &= 1 - 4(x + a) + 5(x^2 + 2xa + a^2) \\ &\quad + 3(x^4 + 4x^3a + 6x^2a^2 + 4xa^3 + a^4) \end{aligned}$$

$$\begin{aligned} &= (1 - 4a + 5a^2 + 3a^4) \\ &\quad + (-4 + 10a + 12a^3)x \\ &\quad + (5 + 18a^2)x^2 \\ &\quad + (12a)x^3 \\ &\quad + 3x^4. \end{aligned}$$

In the last step, we collected like powers of x. Some people say this as "we wrote the answer as a polynomial in x with coefficients that are polynomials in a."

You could obtain this from the TI-89 CAS by entering

```
Expand[1 -4(x + a) + 5(x + a)^2 + 3(x + a)^4,x]
```

In *Mathematica*, you'd type:

```
Collect[Expand[1 -4(x + a) + 5(x + a)^2 + 3(x + a)^4],x]
```

Let's work out $f(x+a)$ for the generic polynomial above, and we'll write the answer as a polynomial in x with coefficients that are polynomials in a. Substitution yields

$$f(x+a) = c_0 + c_1(x+a) + c_2(x+a)^2 + \cdots + c_m(x+a)^m$$

by expanding each binomial (using the binomial theorem) and gathering like terms. It's tedious (but easy on a CAS for particular m). You get:

Try it. It's like doing sit-ups. If you do sit-ups with a CAS (*Mathematica*, say), you might try this:

```
f[x_] = r * x^4 + s * x^3
        + t * x^2 + u * x + v
Collect[Expand[f[x + a]], x]
```

$$\begin{aligned} f(x+a) &= (c_0 + c_1 a + c_2 a^2 + \cdots + c_m a^m) \\ &\quad + (c_1 + 2c_2 a + 3c_3 a^2 + \cdots + m c_m a^{m-1})x \\ &\quad + \left(2c_2 + 3 \cdot 2c_3 a \cdots + m(m-1)c_m a^{m-2}\right)\frac{x^2}{2} \\ &\quad + \left(3 \cdot 2c_3 + 4 \cdot 3 \cdot 2c_4 a \cdots + m(m-1)(m-2)c_m a^{m-3}\right)\frac{x^3}{6} \\ &\quad + \\ &\quad \vdots \\ &\quad + (m! c_m)\frac{x^m}{m!}. \end{aligned}$$

Oh my. Look at the "constant term"—the part with no x in it. It's just $f(a)$. And what about the coefficient of x? It is none other than the derivative of f evaluated at a, that is $f'(a)$ (see the *Ways to think about it* below). And the coefficient of x^2? It's $\frac{1}{2}$ of the second derivative (the derivative of the derivative) of f evaluated at a: it's $f''(a)/2$. In fact, if we let $f^{(k)}(x)$ be the kth derivative of f, we have a very nice theorem:

Theorem 3. *If f is a polynomial of degree m,*

$$f(x+a) = f(a) + f'(a)x + \frac{f''(a)}{2!}x^2 + \cdots + \frac{f^{(m)}(a)}{m!}x^m.$$

The "Δ analogue":

$$\begin{aligned} f(x+a) = f(a) &+ \Delta(f)(a)\binom{x}{1} \\ &+ \Delta^2(f)(a)\binom{x}{2} + \cdots \\ &+ \Delta^{(m)}(f)(a)\binom{x}{m} \end{aligned}$$

Can you prove it? See problem 69.

Ways to think about it

You may remember the derivative from calculus. Starting with the graph of a function, $y = f(x)$, the object was to produce a formula for the slope of the tangent to the graph at any point along the graph, if such a slope existed. Equivalently, you were looking for the instantaneous rate of change of the function at any point in its domain (well, at any point for which this rate makes sense). With this motivation, the derivative of f at some number a was defined as the number $f'(a)$ given by

$$\lim_{h \to 0} \frac{f(a+h) - f(a)}{h}.$$

After a fair amount of work, it turned out that for a *polynomial* function f:

$$f(x) = c_0 + c_1 x + c_2 x^2 + \cdots + c_m x^m,$$

you get a nice formula

$$f'(a) = c_1 + 2c_2 a + 3c_3 a^2 + \cdots + m c_m a^{m-1}.$$

We've stumbled upon the derivative here in another, purely algebraic, context. If you calculate $f(x+a)$ as a polynomial in x with coefficients which are polynomials in a, these coefficients turn out to be the successive derivatives of f evaluated at a (and divided by a factorial). So, there's something about the derivative that has nothing to do with limits; if you didn't know any calculus, you'd bump into the formula for $f'(a)$ just by looking at the expansion of $f(x+a)$. As we'll see, this algebraic perspective on derivatives will be a useful tool in analyzing polynomials.

In abstract algebra, there's a whole theory of "formal derivatives" that exploits the properties of a polynomial that can be obtained from its successive derivatives with no recourse to calculus. It's very much like the theory of successive differences developed in Chapter 1.

Put $a = 0$ in Theorem 3 to obtain

Corollary 5. *If f is a polynomial of degree m,*

$$f(x) = f(0) + f'(0)x + \frac{f''(0)}{2!}x^2 + \cdots + \frac{f^{(m)}(0)}{m!}x^m.$$

This gives the interpretation of the coefficients of $f(x)$ when it is written in terms of powers of x:

$$\begin{aligned} c_0 &= f(0) \\ c_1 &= f'(0) \\ c_2 &= \frac{f''(0)}{2!} \\ \vdots \quad & \quad \vdots \end{aligned}$$

$$c_k = \frac{f^{(k)}(0)}{k!}$$

$$\vdots \quad \vdots$$

Example: This answers the question we posed on page 54. Indeed, if

$$f(x) = \frac{2}{3}x^3 - \frac{17}{2}x^2 + \frac{77}{6}x - 6,$$

we now know that

$$\begin{aligned} -6 &= f(0), \\ \frac{77}{6} &= f'(0), \\ -\frac{17}{2} &= \frac{f''(0)}{2}, \\ \frac{2}{3} &= \frac{f'''(0)}{6}. \end{aligned}$$

Our identity

$$f(x) = f(0) + f'(0)x + \frac{f''(0)}{2!}x^2 + \cdots + \frac{f^{(m)}(0)}{m!}x^m$$

is a version of what's known as Taylor's formula. In calculus, it's proved using analytic techniques. Here (remember, these are polynomial functions), it is an *algebraic identity*.

"Analytic techniques" involve using ideas of distance, absolute value, and closeness in the real numbers. And it's proved for functions that aren't necessarily polynomials (like cosine). We got it for polynomials just by calculating.

We can make the analogy between our Δ formula (Theorem 9 on page 37 of Chapter 1) and Taylor more astounding if we rewrite the former a little. It says

$$f(x) = f(0) + \Delta f(0)\binom{x}{1} + \Delta^2 f(0)\binom{x}{2} + \Delta^3 f(0)\binom{x}{3} + \Delta^4 f(0)\binom{x}{4} + \cdots + \Delta^m f(0)\binom{x}{m}.$$

Rearrange the form of the right-hand side by replacing $\binom{x}{k}$ by its definition and then moving the denominator to the "Δ factor":

$$\begin{aligned} f(x) = f(0) + \Delta f(0)x + \Delta^2 f(0)\frac{x(x-1)}{2} + \Delta^3 f(0)\frac{x(x-1)(x-2)}{6} \\ + \Delta^4 f(0)\frac{x(x-1)(x-2)(x-3)}{24} \\ + \cdots + \Delta^m f(0)\frac{x(x-1)(x-2)(x-3)\cdots(x-m+1)}{m!} \end{aligned}$$

so . . .

$$f(x) = f(0) + \Delta f(0)x + \frac{\Delta^2 f(0)}{2}x(x-1) + \frac{\Delta^3 f(0)}{6}x(x-1)(x-2)$$
$$+ \frac{\Delta^4 f(0)}{24}x(x-1)(x-2)(x-3)$$
$$+ \cdots + \frac{\Delta^m f(0)}{m!}x(x-1)(x-2)(x-3)\cdots(x-m+1).$$

Assume f has degree m.

Then, if we define the "falling power" $x^{\underline{p}}$ by

$$x^{\underline{p}} = x(x-1)(x-2)(x-3)\cdots(x-p+1),$$

This notion of "falling power" goes back at least to the 19th century in a book by George Boole of Boolean algebra fame [4].

then our identity becomes

$$f(x) = f(0) + \Delta f(0)x + \frac{\Delta^2 f(0)}{2!}x^{\underline{2}} + \frac{\Delta^3 f(0)}{3!}x^{\underline{3}} + \frac{\Delta^4 f(0)}{4!}x^{\underline{4}}$$
$$+ \cdots + \frac{\Delta^m f(0)}{m!}x^{\underline{m}}.$$

So, we see an analogy emerging.

Ways to think about it

In hindsight, the result of Corollary 5 seems almost too easy. For example, look at

$$f(x) = 7x^4 - 8x^3 + 6x^2 + 9x - 11.$$

Let's "differentiate it down:"

$$f(x) = 7x^4 - 8x^3 + 6x^2 + 9x - 11$$
$$f'(x) = 4 \cdot 7x^3 - 3 \cdot 8x^2 + 2 \cdot 6x + 1 \cdot 9$$
$$f''(x) = 3 \cdot 4 \cdot 7x^2 - 2 \cdot 3 \cdot 8x + 1 \cdot 2 \cdot 6$$
$$f'''(x) = 2 \cdot 3 \cdot 4 \cdot 7x - 1 \cdot 2 \cdot 3 \cdot 8$$
$$f''''(x) = 1 \cdot 2 \cdot 3 \cdot 4 \cdot 7$$
$$f'''''(x) = 0$$

Now, evaluating a polynomial at 0 just produces its constant term, so:

$$f(0) = \mathbf{-11} \quad (0! \cdot \text{ the constant term of } f)$$
$$f'(0) = 1 \cdot \mathbf{9} \quad (1! \cdot \text{ the coefficient of } x)$$
$$f''(0) = 1 \cdot 2 \cdot \mathbf{6} \quad (2! \cdot \text{ the coefficient of } x^2)$$
$$f'''(0) = 1 \cdot 2 \cdot 3 \cdot \mathbf{-8} \quad (3! \cdot \text{ the coefficient of } x^3)$$
$$f''''(0) = 1 \cdot 2 \cdot 3 \cdot 4 \cdot \mathbf{7} \quad (4! \cdot \text{ the coefficient of } x^4)$$

(*continued*)

And this idea works in general (problem 46), providing another proof of Corollary 5. So, what's the big deal? There are a few answers:

- Theorem 3 is more general than the result of Corollary 5; it's a big deal in its own right—a tool that will often come in handy (see below).
- Imagine you didn't already know about derivatives (put yourself in the place of precalculus students, say); you wouldn't know to "differentiate f down" as we did above.
- Our approach shows that the derivative arises in a natural *algebraic* context, with no recourse to calculus.

The next corollary (a generalization of Corollary 5) is an extremely useful consequence of Theorem 3. We get it if we replace x by $x - a$ in Theorem 3:

Corollary 6. *If f is a polynomial of degree m,*

$$f(x) = f(a) + f'(a)(x-a) + \frac{f''(a)}{2!}(x-a)^2 + \cdots + \frac{f^{(m)}(a)}{m!}(x-a)^m.$$

This is called "the Taylor expansion of f about a." The result of Corollary 5 is thus the Taylor expansion of f about 0.

Here's an example of how Corollary 6 is used.

Example: [The Product Rule for Derivatives] Suppose f and g are polynomials. You may remember from calculus that there's a relationship between the derivatives of f, g, and the product fg. This, too, turns out to be a purely algebraic result. One way to prove it is to take two generic expressions ($a_n x^n + \cdots$ and $b_m x^m + \cdots$), multiply them together, and take the formal derivative of the result. That should work, though it would be quite tedious, and, if you didn't know the result in the first place, it might be hard to spot. Here's another approach that makes use of the theorems we have so far.

You certainly *do* remember this if you *teach* calculus.

Let a be a number. Then Corollary 6 can be applied to f, g, and fg to obtain:

$$f(x) = f(a) + f'(a)(x-a) + \frac{f''(a)}{2!}(x-a)^2 + \cdots$$

$$g(x) = g(a) + g'(a)(x-a) + \frac{g''(a)}{2!}(x-a)^2 + \cdots$$

$$(fg)(x) = (fg)(a) + (fg)'(a)(x-a) + \frac{(fg)''(a)}{2!}(x-a)^2 + \cdots$$

$(fg)(x) = f(x)g(x)$ is the *definition* of what we mean by fg. Actually, it's two definitions. If you think "form" it says that to get the polynomial fg, multiply the polynomials f and g together formally. If you "think function," it says that the function fg is defined by the rule $x \mapsto f(x)g(x)$ (and its domain is the same as that of f and g—all of $\mathbb{R}$). Which should you use? Take your pick.

Now, the product of the right-hand sides of the first two equations is the same as the right-hand side of the last equation (because $(fg)(x) = f(x)g(x)$). But $(fg)'(a)$ is the coefficient of $(x-a)$ in the last equation, so the idea is to pick out the coefficient of $(x-a)$ in the product of the first two. Since there's only one way to write a polynomial as powers of $(x-a)$ (problem 68 on page 37 of Chapter 1), we'll know that these two coefficients must be the same. So, where can an $(x-a)$ come from in the product of the first two lines? Certainly not by involving $(x-a)^2$ or any higher power. So, it can only come from

$$\left[f(a)+f'(a)(x-a)\right]\left[g(a)+g'(a)(x-a)\right],$$

which multiplies out to

$$f(a)g(a)+\left[f'(a)g(a)+f(a)g'(a)\right](x-a)+\text{ (some number) }(x-a)^2.$$

Comparing this with the expression for $(fg)'(x)$, we see that $(fg)(a)=f(a)g(a)$ (no surprise) and that

$$(fg)'(a)=f'(a)g(a)+f(a)g'(a).$$

And now the punchline: Both $(fg)'(x)$ and $f'(x)g(x)+f(x)g'(x)$ are polynomials, and a is *arbitrary*, so the polynomials agree for infinitely many inputs. It follows by the "function implies form theorem" (Corollary 3 on page 65) that they are identical:

$$(fg)'(x)=f'(x)g(x)+f(x)g'(x)$$

or, in more compact form:

$$(fg)'=f'g+fg'.$$

Problems

46. Prove Corollary 5 using the method of the *Ways to think about it* on page 72.

47. Let $f(x)=5x^3-3x^2+2x-9$. Write f as a linear combination of powers of $(x-1)$.

Hint for problem 48: Show that $g(x-2)=f(x)$ and use corollary 6.

48. Let $f(x)=x^3-6x^2+3x+10$. Let $g(x)$ be the (polynomial) function whose graph is obtained from the graph of f by shifting it two units to the left. Express g as a polynomial in x and explain how the coefficients in this expression can be obtained from f and its derivatives.

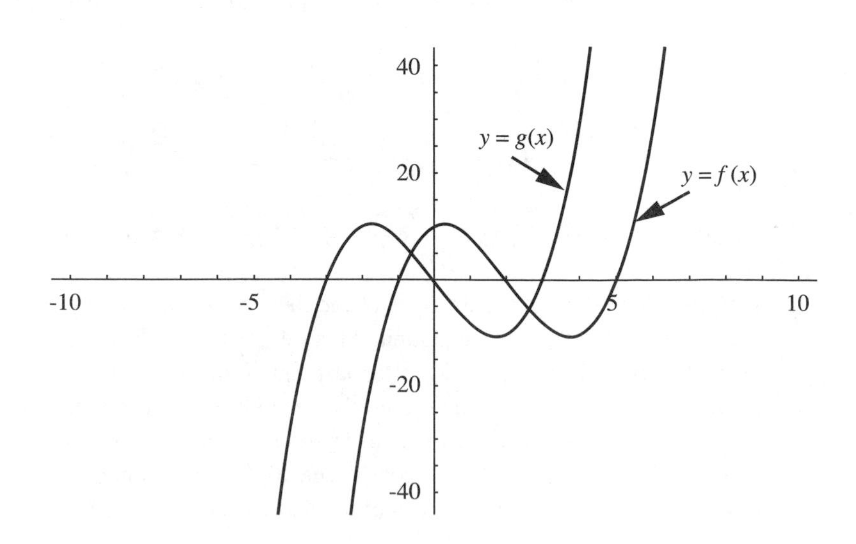

49. Let f be a polynomial function, and let g be the (polynomial) function whose graph is obtained from the graph of f by shifting it a units to the left (where $a > 0$ is a number). Express g as a polynomial in x, expressing its coefficients as expressions obtained from f and its derivatives.

50. Let $f(x) = 5x^3 - 3x^2 + 2x - 9$. Overlay the graph of f with the graphs of

$$f_r(x) = \sum_{k=0}^{r} \frac{f^{(k)}(0)}{k!} x^k$$

for $r = 1$ to 3. Describe the resulting picture.

51. If f is a polynomial, show that

In what sense is this sum infinite?

$$f(x) = \sum_{k=0}^{\infty} \frac{f^{(k)}(a)}{k!} (x - a)^k.$$

52. Use Corollary 6 to prove Theorem 3.

Problem 52 shows that Corollary 6 and Theorem 3 are *equivalent*: either implies the other.

53. Use Corollary 6 to prove the remainder theorem.

54. Revisit problem 43 on page 68, where you found an explicit formula for the quotient when $f(x)$ is divided by $x - a$. Use Corollary 6 to express this quotient in terms of the derivatives of f.

55. Suppose f is a polynomial. Show that $f(x)$ is divisible by $(x - a)^2$ if and only if $f(a) = f'(a) = 0$.

Hint: Use problem 54.

56. Suppose f is a polynomial. Show that $f(x)$ is divisible by $(x - a)^k$ if and only if $f^{(j)}(a) = 0$ for all $0 \le j < k$. Here, $f^{(j)}$ is the jth derivative of f.

57. Show that if f and g are polynomials and c is a number,

(a) $(f + g)' = f' + g'$, **(b)** $(cf)' = cf'$.

Problem 57 isn't so bad by calculation with generic forms; the method of the example on page 73 also works. So, in the language of Chapter 1, the problem says that the derivative is *linear*.

58. Suppose that f and g are polynomials and $f'(x) = g'(x)$. Show that f and g differ by a constant.

59. Show that if $f'(x) = x^k$ for some integer k, then there is a constant C so that

$$f(x) = \frac{1}{k+1} x^{k+1} + C.$$

60. Classify all polynomials f so that $f'(x) = x^4 + 3x^2 + 2x + 1$.

61. Classify all polynomials f so that $f'(x) = 0$.

62. Let $f(x) = (x - \alpha)(x - \beta)(x - \gamma)$ where α, β, and γ are numbers. Show that:

Hint: Use the product rule for derivatives (see page 73).

$$f'(x) = \frac{f(x)}{x - \alpha} + \frac{f(x)}{x - \beta} + \frac{f(x)}{x - \gamma}.$$

63. Suppose f and g are polynomials. Use the method of the example on page 73 to find an expression for $(f \circ g)'$ in terms of f, g, f', and g'.

Hints: Suppose a is any number. Corollary 6 can be applied to $f \circ g$ to obtain:

$$(f \circ g)(x) = (f \circ g)(a) + (f \circ g)'(a)(x - a) + \frac{(f \circ g)''(a)}{2}(x - a)^2.$$

We can also apply the same corollary to f, this time replacing x by $g(x)$ and a by $g(a)$ to obtain

$$f\big(g(x)\big) = f\big(g(a)\big) + f'\big(g(a)\big)\big(g(x) - g(a)\big) + \frac{f''\big(g(a)\big)}{2}\big(g(x) - g(a)\big)^2 + \cdots \qquad (*)$$

Since $(f \circ g)(x) = f\big(g(x)\big)$, the right-hand sides of these expressions are equal. Now expand $g(x)$ in powers of $(x - a)$, replace it by this expansion in $(*)$, and compare the coefficients of $(x - a)$.

The difference of the derivative is the derivative of the difference.

64. Show that if f is a polynomial,

$$\Delta(f') = (\Delta f)'.$$

Problems 65–68 look at the connection between the results of this section and the calculus of polynomial functions.

65. Suppose $f(x)$ is a polynomial function defined by

$$f(x) = c_0 + c_1 x + c_2 x^2 + \cdots + c_n x^n$$

and a is a real number. Show that

$$\lim_{h \to 0} \frac{f(a + h) - f(a)}{h} = c_1 + c_2 a + \cdots + c_n a^{n-1}$$

where f' is the derivative of f, as we have defined it in this chapter.

Hint: First of all, be sure about what the problem says. The equation above is taken as the definition of the derivative in calculus courses. Forget about that for now. Here, we already have an algebraic definition of derivative (the one on page 69), and we're trying to show that the limit on the left-hand side of the equation produces precisely $f'(a)$.

Use the identity in Theorem 3 (with the right choice of substitution) to show that

$$\frac{f(h + a) - f(a)}{h} = f'(a) + \frac{f''(a)}{2}h + \frac{f'''(a)}{6}h^2 + \frac{f''''(a)}{4!}h^3 + \cdots$$

So, h is a factor of $f(h + a) - f(a)$, and the quotient is the polynomial in h on the right of the above equation. Since polynomials are continuous everywhere, the limit of the right-hand side when h goes to 0 is just the value of this expression when h *equals* 0.

66. Explain why, if f is a polynomial and a is a number, $f'(a)$ is a good way to define the slope of the line tangent to the graph of f at $x = a$.

67. Show that, if f is a polynomial function, $f'(x)$ can be obtained as a limit:

$$f'(x) = \lim_{h \to 0} \frac{f(x+h) - f(x)}{h}.$$

By problem 65, this equation holds for any number $a, \ldots.$

68. Show that if f is a polynomial, the coefficient of x in $f(x)$ is the slope of the tangent to the graph of f at $x = 0$. In addition to an algebraic argument, can you give a geometric argument for why this must be true?

69. Prove the discrete analogue of Theorem 3:

$$f(x+a) = f(a) + \Delta(f)(a)\binom{x}{1} + \Delta^2(f)(a)\binom{x}{2} + \cdots + \Delta^{(m)}(f)(a)\binom{x}{m}. \qquad (*)$$

Hint: One way to think about this is to think of a as a fixed number. Then, mimicking the "up and over" approach of Chapter 1 (see page 13 of Chapter 1), supposing that n is an integer with $0 \le n \le m$, calculate like this:

$$\begin{aligned} f(a+n) &= f(a+n-1) + \Delta(f)(a+n-1) \\ &= f(a+n-2) + 2\Delta(f)(a+n-2) + \Delta^2(f)(a+n-2) \\ &= f(a+n-3) + \binom{3}{1}\Delta(f)(a+n-3) \\ &\quad + \binom{3}{2}\Delta^2(f)(a+n-3) + \Delta^3(f)(a+n-3) \\ &\;\;\vdots \\ &= f(a) + \binom{n}{1}\Delta(f)(a) + \binom{n}{2}\Delta^2(f)(a) + \cdots + \Delta^n(f)(a). \end{aligned}$$

The nth step in this calculation looks just like $(*)$ with x replaced by n, so the polynomials on each side of $(*)$ agree for $m+1$ inputs . . . .

Problems 70–74 give numerical previews for some connections between the derivative and Δ. More details are in the next section.

70. Suppose that $f(x) = x^5 - 3x^4 - 2x^3 + 2x^2 - 5x + 1$. Show that

$$f'(x) = \Delta(f)(x) - \frac{\Delta^2(f)(x)}{2} + \frac{\Delta^3(f)(x)}{3} - \frac{\Delta^4(f)(x)}{4} + \frac{\Delta^5(f)(x)}{5}.$$

For problems 70–73, first calculate each of the $\Delta^k(f)$ for $1 \le k \le 5$. A CAS will help.

71. Suppose that $f(x) = x^5 - 3x^4 - 2x^3 + 2x^2 - 5x + 1$. Show that

$$f''(x) = \Delta^2(f)(x) - \Delta^3(f)(x) + \frac{11\Delta^4(f)(x)}{12} - \frac{5\Delta^5(f)(x)}{6}.$$

72. Suppose that $f(x) = x^5 - 3x^4 - 2x^3 + 2x^2 - 5x + 1$. Show that

$$f'''(x) = \Delta^3(f)(x) - \frac{3\Delta^4(f)(x)}{2} + \frac{7\Delta^5(f)(x)}{4}.$$

73. Suppose that $f(x) = x^5 - 3x^4 - 2x^3 + 2x^2 - 5x + 1$. Show that

$$f''''(x) = \Delta^4(f)(x) - 2\Delta^5(f)(x) \quad \text{and that} \quad f'''''(x) = \Delta^5(f)(x).$$

Of course, don't pick the same f as the one in problems 70–73.

74. Pick any polynomial f (of degree at most 5) that you like. Show that the results of problems 70–73 hold without modification for this f.

2.4 Up a Level

Most of the results in this chapter hold (in some form) for non-polynomial functions as well. See [4] and [11] for more details.

More about symmetric functions in section 2.6.

Problems 70–74 hinted that there might be a way to relate the derivatives of a (polynomial) function f to the functions $\Delta^k(f)$. Let's make this more precise. The key tools are problems 67 and 69.

Notice that the identity in problem 69 is "symmetric" in x and a. In other words, instead of writing it as

$$f(x+a) = f(a) + \Delta(f)(a)\binom{x}{1} + \Delta^2(f)(a)\binom{x}{2} + \cdots + \Delta^{(m)}(f)(a)\binom{x}{m},$$

we could switch the roles of x and a and write it like this:

$$f(x+a) = f(x) + \Delta(f)(x)\binom{a}{1} + \Delta^2(f)(x)\binom{a}{2} + \cdots + \Delta^{(m)}(f)(x)\binom{a}{m}.$$

Now replace each $\binom{a}{k}$ by its polynomial form:

$$\begin{aligned} f(x+a) = f(x) &+ \Delta(f)(x)a + \Delta^2(f)(x)\frac{a(a-1)}{2} \\ &+ \Delta^3(f)(x)\frac{a(a-1)(a-2)}{3!} \\ &\cdots + \Delta^{(m)}(f)(x)\frac{a(a-1)(a-2)\cdots(a-m+1)}{m!} \end{aligned}$$

You'll be asked to fill in the details in problem 75.

Replace a by h, combine this identity with problem 67, and do some rearranging of terms, and you'll get a nice theorem:

For polynomial functions, this sum is actually finite.

Theorem 4. *If f is a polynomial function,*

$$\begin{aligned} f'(x) &= \Delta(f)(x) - \frac{\Delta^2(f)(x)}{2} + \frac{\Delta^3(f)(x)}{3} \\ &\quad - \frac{\Delta^4(f)(x)}{4} + \frac{\Delta^5(f)(x)}{5} - \cdots \\ &= \sum_{k=1}^{\infty}(-1)^{k-1}\frac{\Delta^k(f)(x)}{k}. \end{aligned}$$

Now, as previewed in problem 74, it turns out that the results of problems 70–73 have nothing to do with a particular f. They hold for any polynomial of degree 5 or less. And there are similar theorems for polynomials of higher degree. In fact, there's a coherent theory that ties all this together. Here's the briefest sketch of how it works.

Look at Theorem 4 again. It says that, if f is any polynomial,

$$f'(x) = \Delta(f)(x) - \frac{\Delta^2(f)(x)}{2} + \frac{\Delta^3(f)(x)}{3} - \frac{\Delta^4(f)(x)}{4} + \frac{\Delta^5(f)(x)}{5} - \cdots$$

Actually, this is "up a notch" from the discussion on page 56: The addition of functions is still defined by the addition of their outputs, but in this case, the outputs of the "functions" being added (the Δ^k/k) are *themselves* functions (or polynomials, depending on your persuasion). So we're adding functions that are defined on nonnumerical creatures. Abstractions often build like this in mathematics.

We could write this more compactly as

$$f' = \left(\Delta - \frac{\Delta^2}{2} + \frac{\Delta^3}{3} - \frac{\Delta^4}{4} + \frac{\Delta^5}{5} - \cdots\right)(f)$$

where it's understood that the sum in the first parentheses is the sum of functions (as described in *Ways to think about it* on page 56) and, for any particular polynomial f, the sum is finite. Now, to get f'', we could think like this:

$$\begin{aligned} f'' &= (f')' \\ &= \left(\left(\Delta - \frac{\Delta^2}{2} + \frac{\Delta^3}{3} - \frac{\Delta^4}{4} + \frac{\Delta^5}{5} - \cdots\right)(f)\right)' \\ &= \left(\Delta - \frac{\Delta^2}{2} + \frac{\Delta^3}{3} - \frac{\Delta^4}{4} + \frac{\Delta^5}{5} - \cdots\right) \\ &\quad \times \left\{\left(\Delta - \frac{\Delta^2}{2} + \frac{\Delta^3}{3} - \frac{\Delta^4}{4} + \frac{\Delta^5}{5} - \cdots\right)(f)\right\} \\ &= \left(\left(\Delta - \frac{\Delta^2}{2} + \frac{\Delta^3}{3} - \frac{\Delta^4}{4} + \frac{\Delta^5}{5} - \cdots\right)\right. \\ &\quad \left.\circ \left(\Delta - \frac{\Delta^2}{2} + \frac{\Delta^3}{3} - \frac{\Delta^4}{4} + \frac{\Delta^5}{5} - \cdots\right)\right)(f). \end{aligned}$$

And, by problem 12 on page 60 (suitably generalized), we can operate on the two big parentheses "algebra style" and multiply them out as if $+$ and $\circ$ with these Δ expressions behaved like $+$ and $\times$ with ordinary (algebra 2) expressions, getting

$$f'' = \left(\Delta^2 - \Delta^3 + \frac{11\Delta^4}{12} - \frac{5\Delta^5}{6} + \frac{137\Delta^6}{180} - \frac{7\Delta^7}{10} + \cdots\right)(f).$$

The "$\cdots$" here doesn't mean that there's an obvious pattern, just that you could keep going with the multiplication.

Since the f of problem 71 has degree 5, we could drop all the Δ expressions after Δ^5.

So, the problem of converting from Δ to derivative comes down to formal algebra again: To get the kth derivative, we need to compute formal powers

$$\left(\Delta - \frac{\Delta^2}{2} + \frac{\Delta^3}{3} - \frac{\Delta^4}{4} + \frac{\Delta^5}{5} - \cdots\right)^k$$

and go out far enough (depending on the degree of the polynomial). We've now let the meaning of Δ fade into the background, and we're calculating with formal expressions in Δ.

And just look at the form of the thing we're raising to a power:

$$\Delta - \frac{\Delta^2}{2} + \frac{\Delta^3}{3} - \frac{\Delta^4}{4} + \frac{\Delta^5}{5} - \cdots.$$

Here, log means ln.

Look familiar? Maybe not, if you've been away from calculus for a few years, but it turns out that this is the formal power series for the natural logarithm of $1 + \Delta$. Indeed, if you ask a CAS for the series for $\log(1 + \Delta)$ (with no meaning attached to Δ, of course—you could use any symbol instead of it), you get

$$\log(1 + \Delta) = \Delta - \frac{\Delta^2}{2} + \frac{\Delta^3}{3} - \frac{\Delta^4}{4} + \frac{\Delta^5}{5} - \cdots.$$

Ah, but we could just as well ask the CAS for the series for $\left(\log(1+\Delta)\right)^k$ for various k, saving a great deal of algebraic calculation. We asked *Mathematica* to do just this, and it produced the following table:

The exact command that produced this is

```
TableForm[Table[
{n,
 (Series[Log[1 + V],
    {V, 0, 6}])^k },
      {k, 1, 5}]]
```

"V" was used as the indeterminate and then changed to Δ in the table to the left.

k	the first 6 terms of the series for $\left(\log(1 + \Delta)\right)^k$
1	$\Delta - \frac{\Delta^2}{2} + \frac{\Delta^3}{3} - \frac{\Delta^4}{4} + \frac{\Delta^5}{5} - \frac{\Delta^6}{6} + \cdots$
2	$\Delta^2 - \Delta^3 + \frac{11\Delta^4}{12} - \frac{5\Delta^5}{6} + \frac{137\Delta^6}{180} - \frac{7\Delta^7}{10} + \cdots$
3	$\Delta^3 - \frac{3\Delta^4}{2} + \frac{7\Delta^5}{4} - \frac{15\Delta^6}{8} + \frac{29\Delta^7}{15} - \frac{469\Delta^8}{240} + \cdots$
4	$\Delta^4 - 2\Delta^5 + \frac{17\Delta^6}{6} - \frac{7\Delta^7}{2} + \frac{967\Delta^8}{240} - \frac{89\Delta^9}{20} + \cdots$
5	$\Delta^5 - \frac{5\Delta^6}{2} + \frac{25\Delta^7}{6} - \frac{35\Delta^8}{6} + \frac{1069\Delta^9}{144} - \frac{285\Delta^{10}}{32} + \cdots$

The spiral of abstractions here is getting pretty steep, and we've left out many details, but the point, in addition to the surprising mathematical connections, is that sometimes you get a lot of mileage by ignoring the "meaning" of the variable and concentrating on how you can calculate with it in a formal way. By the way, see any patterns in the coefficients in this table?

Problems

75. Use the identity on page 78 and problem 67 to prove Theorem 4.

76. Here's a difference table for a function f:

Input	Output	Δ	Δ^2	Δ^3	Δ^4	Δ^5
0	−5	−1	14	132	240	120
1	−6	13	146	372	360	
2	7	159	518	732		
3	166	677	1250			
4	843	1927				
5	2770					

The table is from problem 17 on page 21 of Chapter 1.

Without finding a formula for $f(x)$, find

(a) $f'(0)$ **(b)** $f''(0)$ **(c)** $f'''(0)$
(d) $f^{(4)}(0)$ **(e)** $f^{(5)}(0)$ **(f)** $f'(2)$
(g) $f''(3)$ **(h)** $f'''(2)$ **(i)** $f^{(4)}(1)$

77. Find a formula for the function tabulated in problem 76 using the results of that problem and Corollary 5 on page 70.

The table on page 80 should be useful, too.

78. Let's push this formal algebra a little further. Suppose we let D stand for the derivative operator, so that $D(f) = f'$. So, we have

$$\log[1 + \Delta] = D.$$

This is just so that we can use ordinary functional notation for the derivative: $D(f)$ instead of f'. If we were "thinking calculus," we could have used d/dx instead of D.

Now, forget the meaning of the symbols. If this were ordinary algebra, we could solve for Δ, obtaining

$$\Delta = \exp(D) - 1$$

where exp is the exponential function—the inverse of the natural logarithm. But exp has a formal power series, too:

See [15], Chapter 6, for the connections between exp and log.

$$\exp(D) = 1 + D + \frac{D^2}{2!} + \frac{D^3}{3!} + \frac{D^4}{4!} + \cdots$$

so

$$\exp(D) - 1 = D + \frac{D^2}{2!} + \frac{D^3}{3!} + \frac{D^4}{4!} + \cdots.$$

Well, if the formalism holds, we should have an identity:

$$\begin{aligned}\Delta(f) &= \left(D + \frac{D^2}{2!} + \frac{D^3}{3!} + \frac{D^4}{4!} + \cdots\right)(f)\\ &= f' + \frac{f''}{2!} + \frac{f'''}{3!} + \frac{f''''}{4!} + \cdots.\end{aligned}$$

Show that this is, in fact, a valid identity if

$$f(x) = x^5 - 3x^4 - 2x^3 + 2x^2 - 5x + 1.$$

79. Prove that the identity in problem 78 holds in general; that is, prove the following theorem:

Theorem 5. *If f is a polynomial function, then*

$$\Delta(f) = f' + \frac{f''}{2!} + \frac{f'''}{3!} + \frac{f''''}{4!} + \cdots.$$

Hint: By definition,

$$\Delta(f)(x) = f(x+1) - f(x).$$

Now put $x = 1$ and $a = x$ in the result of Theorem 3 on page 69.

80. Suppose f is a polynomial. Establish the identity:

$$f(x+a+1) - f(x+a) = f'(x) + \left((a+1)^2 - a^2\right)\frac{f''(x)}{2!} + \left((a+1)^3 - a^3\right)\frac{f'''(x)}{3!} + \cdots.$$

Hint: Do *not* use Theorem 5. Instead, use Theorem 3.

2.5 Transformations

Polynomials (forms and functions) arose in the context of people trying to solve polynomial *equations*: equations like $5x^4 - 3x^3 + 2x + 1 = 0$. In modern terminology, they were looking for the zeros of polynomial functions. For equations of small degree, there are *formulas* that allow you to find the roots in a purely mechanical way.

- For equations of degree 1, like $ax + b = 0$ ($a \neq 0$), you just "solve for x" as in algebra 1:

$$x = -\frac{b}{a}.$$

If you did problem 26 on page 61, you proved the quadratic formula—if you teach algebra, you probably know the proof quite well.

- For equations of degree 2 like $ax^2 + bc + c = 0$ ($a \neq 0$), you have the quadratic formula

$$x = \frac{-b + \sqrt{b^2 - 4ac}}{2a} \quad \text{or} \quad x = \frac{-b - \sqrt{b^2 - 4ac}}{2a}.$$

- There's a similar formula, known as "Cardan's formula," for polynomials of degree 3 (we'll look at it in problems 88–91 and again in the next chapter) and there's even one for polynomials of degree 4 (see [2], for example).

- All this was known well before 1600. But the search went on for similar formulas for higher degrees until 1820 or so when the Norwegian mathematician Abel showed (at the age of 19) that no such formula was possible for degree higher than 4. It's certainly possible to solve *some* equations of degree 5 or higher, and there are even formulas for solving whole classes of them (like equations of the form $x^n - 1 = 0$, where n is a positive integer). Furthermore, there are methods to approximate the roots of an equation to any desired degree of accuracy, and there are even exact methods involving fancy analysis. But that's not the game algebraists were playing; they wanted a general *algebraic* method that handles every equation of a given degree. And Abel showed that this is impossible.

There's the term "algebraic" again; here it means that there's no formula like the quadratic formula (involving only the four operations of arithmetic and the extraction of roots) that solves, say,

$$a_5x^5 + a_4x^4 + a_3x^3 + a_2x^2 + a_1x + a_0 = 0$$

in terms of the a_i.

So, the main agenda in algebra—the search for algebraic formulas that solve algebraic equations—was shown to be futile. But, as is typical in our discipline, the *methods* used to establish this futility became objects of study in themselves, forming one of the foundations of abstract algebra. Abel, and independently Galois (1811∼1832), showed that the problem of finding a formula for the roots of an equation was connected to the problem of finding certain relationships among the roots and the coefficients. For equations of degree higher than 5, the relationships were so meager that no algebraic formula was possible (we'll say more about this in the next section).

Relationships: for example, in the quadratic equation $x^2 + bx + c = 0$, the sum of the roots is $-b$ and the product is c. That turns out to be enough to find the quadratic formula. Briefly, if the roots of our quadratic equation are α and β, then $\alpha + \beta = -b$ and $\alpha\beta = c$. Since

$$(\alpha - \beta)^2 = (\alpha + \beta)^2 - 4\alpha\beta$$

$\alpha - \beta = \sqrt{b^2 - 4c}$. Add this equation to $\alpha + \beta = -b$ and solve for α.

Attention thus shifted away from finding general formulas to finding ways to tell if special relationships existed among the roots of a given equation. One way to do this revolved around what was called the *transformation of equations:* given an equation, how do you find the equation whose roots are the squares (or negatives, or reciprocals, or. . .) of the roots of the given equation? If any of these "transformed" equations turns out to be easy to solve, you are in business. And, if any of the transformed equations turned out to be the *same* as your original, you have a new piece of information at your disposal (the roots of the original come in reciprocal pairs, for example).

In this section and the next, we'll look at two techniques from this classical part of algebra. The first is based on carefully chosen changes of variable (substitutions), including variants of the Taylor expansions (Corollary 6 on page 73). The second, taken up in section 2.6, uses the result of problem 33 (page 62) that shows how to express the coefficients of an equation in terms of the roots.

Ways to think about it

As usual, our interest in these techniques stems from several features: The mathematics is interesting in its own right, and it laid the foundation for major developments in algebra that inspire research to this day. The *thinking* behind these topics is quite useful for algebra teachers: changing the form of an expression to make it more useful and (in the next section) looking for "hidden" meaning in algebraic forms. The ideas help put school mathematics (completing the square and the sum and product of the roots for quadratics, for example) in a broader perspective. And some of the results are just plain pretty.

Scaling the Roots

Suppose you have a polynomial function f and you want to find a polynomial whose zeros are, say, twice the zeros of f. The logic proceeds like this: Let α be any of the zeros of f (so that $f(\alpha) = 0$). We're searching for a polynomial g so that $g(2\alpha) = 0$. This will work if $g(2x) = f(x)$, in other words, if we let $g(x)$ be $f(\frac{x}{2})$. Then, for any zero α of f,

$$g(2\alpha) = f\left(\frac{2\alpha}{2}\right) = f(\alpha) = 0.$$

We could just find the roots of f, double them, and find an equation whose roots are the doubles. Why is this not a practical method?

Example: Suppose $f(x) = 3x^3 + 5x^2 - x + 7$, and we want an equation whose roots are twice the roots of the equation $3x^3 + 5x^2 - x + 7 = 0$. Let

$$\begin{aligned} g(x) &= f\left(\frac{x}{2}\right) \\ &= 3\left(\frac{x}{2}\right)^3 + 5\left(\frac{x}{2}\right)^2 - \left(\frac{x}{2}\right) + 7. \end{aligned}$$

So, the equation

$$3\left(\frac{x}{2}\right)^3 + 5\left(\frac{x}{2}\right)^2 - \left(\frac{x}{2}\right) + 7 = \frac{3x^3}{8} + \frac{5x^3}{4} - \frac{x}{2} + 7 = 0$$

has roots that are double the roots of the equation $f(x) = 0$. Of course, we can clear fractions (multiply both sides by 8) and get a simpler equation:

$$3x^3 + \mathbf{2} \cdot 5x^3 - \mathbf{4} \cdot x + \mathbf{8} \cdot 7 = 0.$$

All the generalizations of the examples to theorems will be left to problem 81.

This method generalizes:

Theorem 6. *To find an equation whose roots are c times the roots of $f(x) = 0$, put f in normal form and then multiply its coefficients by, respectively, $1, c, c^2, \ldots,$ starting at the highest degree term.*

Reciprocal Roots

This method shouldn't work if one root of the equation is 0. Where does it go wrong?

Suppose you want to find a polynomial whose zeros are the reciprocals of the zeros of a polynomial function f. Let α be any of the zeros of f (so that $f(\alpha) = 0$). We're searching for a polynomial g so that $g(1/\alpha) = 0$. Let $g(x)$ be $f(1/x)$. Then, for any zero α of f,

$$g\left(\frac{1}{\alpha}\right) = f\left(\frac{1}{1/\alpha}\right) = f(\alpha) = 0.$$

The trouble is, g isn't a polynomial. But clearing fractions will do the trick.

Example: Let's find an equation whose roots are the reciprocals of the equation $f(x) = 3x^3 + 5x^2 - x + 7 = 0$. As in the above discussion, let $g(x) = f(1/x)$. The roots of g are the reciprocals of the roots of f. And we have

$$g(x) = 3\left(\frac{1}{x}\right)^3 + 5\left(\frac{1}{x}\right)^2 - \left(\frac{1}{x}\right) + 7$$

so an equation with the desired properties is

$$3\left(\frac{1}{x}\right)^3 + 5\left(\frac{1}{x}\right)^2 - \left(\frac{1}{x}\right) + 7 = \frac{3}{x^3} + \frac{5}{x^2} - \frac{1}{x} + 7 = 0.$$

Multiply both sides by x^3 to get a *polynomial* equation with the desired properties:

$$7x^3 - x^2 + 5x + 3 = 0.$$

The coefficients are the same as those of f, written in reverse order.

This method generalizes:

Theorem 7. *To obtain a polynomial whose zeros are the reciprocals of the zeros of f (assuming $f(0) \neq 0$), form the polynomial whose coefficients are the same as those of f, but written in reverse order.*

Translating the Roots

Suppose we want to find an equation whose roots are a less than the roots of f, where a is some number. Let α be any of the zeros of f. We're searching for a polynomial g so that $g(\alpha - a) = 0$. Corollary 6 makes it easy to calculate the coefficients of g, one at a time. The corollary says that (if f has degree m)

$$f(x) = f(a) + f'(a)(x-a) + \frac{f''(a)}{2!}(x-a)^2 + \cdots + \frac{f^{(m-1)}(a)}{(m-1)!}(x-a)^{m-1} + \frac{f^{(m)}(a)}{m!}(x-a)^m.$$

Simply let

$$g(x) = f(a) + f'(a)x + \frac{f''(a)}{2!}x^2 + \cdots + \frac{f^{(m-1)}(a)}{(m-1)!}x^{m-1} + \frac{f^{(m)}(a)}{m!}x^m.$$

In other words, let $g(x) = f(x+a)$. So,

$$g(\alpha - a) = f(\alpha - a + a) = f(\alpha) = 0.$$

We have a theorem:

Theorem 8. *If f is a polynomial, a is a number, and g is defined by*

$$g(x) = f(a) + f'(a)x + \frac{f''(a)}{2!}x^2 + \cdots + \frac{f^{(m-1)}(a)}{(m-1)!}x^{m-1} + \frac{f^{(m)}(a)}{m!}x^m,$$

then the zeros of g are "a" less than those of f.

Example: *"Completing the Square."* Let $f(x) = 3x^2 - 12x + 1$. Let's translate the roots of f to produce an equation with no "degree 1" term. If $g(x)$ is the function obtained by decreasing the roots of f by a, we have by Theorem 8,

This method can be used to derive the quadratic formula. It can also be used to remove the "penultimate" term from any equation (of any degree).

$$g(x) = f(a) + f'(a)x + \frac{f''(a)}{2}x^2.$$

So, we want $f'(a) = 0$. Since $f'(x) = 6x - 12$, $a = 2$, and

$$g(x) = -11 + 3x^2.$$

Note that it's easy to find the zeros of g; they are $\pm\sqrt{\frac{11}{3}}$, so, the zeros of f are 2 more than these,

$$2+\sqrt{\frac{11}{3}} \qquad 2-\sqrt{\frac{11}{3}}.$$

Problems

81. Generalize the examples in this section to produce proofs of Theorems 6–8.

82. Find an equation whose roots are the negatives of the roots of the equation

$$6x^5+3x^4+9x^3+7x^2+2x+5=0.$$

83. Consider the equation $5x^3-3x^2+4x-7=0$. Find an equation whose roots are a multiple of the roots of this equation, whose coefficients are integers, and whose leading coefficient is 1.

A polynomial with integer coefficients and leading coefficient 1 is called *monic.*

84. Let $f(x)=a_nx^n+a_{n-1}x^{n-1}+\cdots+a_0$, where the coefficients are integers. Explain how to find a new equation whose roots are a multiple of the roots of this one, whose coefficients are integers, and whose leading coefficient is 1.

85. Consider the equation $8x^2-10x-3=0$.

(a) Translate the roots of the equation so that the transformed equation has no x term.

(b) Transform the equation in problem 85a by scaling the roots so that the coefficient of x^2 is 1.

(c) Solve the equation in problem 85b.

(d) Use the result of problem 85c to solve $8x^2-10x-3=0$.

86. Consider the equation $ax^2+bx+c=0$.

(a) Translate the roots of the equation so that the transformed equation has no x term.

(b) Transform the equation in problem 86a by scaling the roots so that the coefficient of x^2 is 1.

(c) Solve the equation in problem 86b.

(d) Use the result of problem 86c to solve $ax^2+bx+c=0$, thus deriving the quadratic formula.

87. Consider the equation $8x^3+36x^2+52x+24=0$.

Here, we want $f''(a)=0$.

(a) Translate the roots of the equation so that the transformed equation has no x^2 term.

(b) Transform the equation in problem 87a by scaling the roots so that the coefficient of x^3 is 1.

(c) Solve the equation in problem 87b.

(d) Use the result of problem 87c to solve $8x^3+36x^2+52x+24=0$.

Problems 88–90 outline a proof of Cardan's formula.

88. Consider the equation $ax^3 + bx^2 + cx + d = 0$.

(a) Translate the roots of the equation so that the transformed equation has no x^2 term.

(b) Transform the equation in problem 88a by scaling the roots so that the coefficient of x^3 is 1.

(c) Explain how, if you were able to solve the equation in problem 88b, you'd be able to solve *any* cubic equation.

89. The equation $x^3 - 6x - 9 = 0$ is the same form as the equation in problem 88c. It can be solved in several ways, but one that's especially useful here comes from the identity of problem 34 on page 62:

$$(r+s)^3 - 3rs(r+s) - (r^3+s^3) = 0.$$

The idea is to find r and s so that

$$3rs = 6 \quad \text{and} \quad r^3 + s^3 = 9.$$

Then $r + s$ will satisfy the equation. Show that this amounts to solving a "quadratic-like" equation, and find r and s that work.

Once you find one root α, you can divide the $x^3 - 6x - 9$ by $x - \alpha$ and find the zeros of the resulting quadratic function.

90. Let's use the method of problem 89 in the general case. By problem 88c, we can get a formula to solve any cubic if we can get a formula to solve a cubic of the form

$$x^3 + px + q = 0.$$

Show that if we represent a root as a sum $r + s$,

$$r^3 + s^3 = -q \quad \text{and} \quad rs = -\frac{p}{3}.$$

Replace s by $-p/3r$ in the first equation and solve the resulting equation for r^3, showing that

$$r^3 = \frac{-q + \sqrt{q^2 + 4(p^3/27)}}{2} \quad \text{and} \quad s^3 = \frac{-q - \sqrt{q^2 + 4(p^3/27)}}{2}.$$

Hence

$$r = \sqrt[3]{\frac{-q + \sqrt{\frac{27q^2+4p^3}{27}}}{2}} \quad \text{and} \quad s = \sqrt[3]{\frac{-q - \sqrt{\frac{27q^2+4p^3}{27}}}{2}},$$

and a root of the cubic is

$$\sqrt[3]{\frac{-q + \sqrt{\frac{27q^2+4p^3}{27}}}{2}} + \sqrt[3]{\frac{-q - \sqrt{\frac{27q^2+4p^3}{27}}}{2}}.$$

This is one version of Cardan's formula.

The cubic $x^3 + px + q$ is said to be "depressed" because its x^2 term has coefficient 0. Maybe it's depressed because it misses it.

There are many details to worry about here. There are choices to make when you take cube roots, and the expressions under the square roots could be negative. But what's here gives the "bones" of the derivation. We'll fill things out a little more in Chapter 3.

91. The equation $x^3 - 4x + 3 = 0$ can be solved by noticing that 1 is a root, so it can be reduced to a quadratic.

(a) Find all the roots of $x^3 - 4x + 3 = 0$.

We'll return to this dilemma in chapter 3.

(b) Use Cardan's formula from problem 90 to solve the same equation. Which root does it produce?

92. Let f be a quadratic polynomial function. Show that transforming f into a quadratic with no x term has the effect of translating the graph of f so that its axis of symmetry is the y-axis.

93. Let f be a cubic polynomial function. Show that transforming f into a cubic with no x^2 term has the effect of translating the graph of f so that its point of symmetry (that is, its inflection point) is on the y-axis.

How many ways are there to do this?

94. Let $f(x) = 2x^3 - 2x^2 - 2x + 3$. Transform f by translation into a function with no x term.

In problem 95, "remove a term" means "make the coefficient of that term 0 by translating the roots."

95. Suppose $f(x) = a_n x^n + a_{n-1}x^{n-1} + \cdots + a_1 x + a_0$.

(a) Show that removing the term of degree $n-1$ requires the solution of a linear equation.

(b) Show that removing the term of degree $n-2$ requires the solution of a quadratic equation.

(c) Show that removing the term of degree $n-k$ requires the solution of an equation of degree k.

(d) Why is removing the constant term equivalent to finding a zero of f?

96. A *reciprocal equation* is an equation with the property that, if α is a root, so is $1/\alpha$.

(a) Show that

$$\frac{x^7-1}{x-1} = 0$$

is a reciprocal equation.

(b) Show that the equation $3x^3 - 4x^2 - 4x + 3 = 0$ is a reciprocal equation.

(c) Find two more reciprocal equations.

(d) Characterize all reciprocal equations in terms of their coefficients.

2.6 Coefficients and Zeros

In section 2.3 and the ones following it, we interpreted the coefficients in a polynomial function in terms of the values of the function and its derivatives. In this final section, we look at another interpretation of the coefficients, this time in terms of the zeros of the function.

If the leading coefficient is not 1, we can do a scale change, as in the last section.

Suppose f is a polynomial of degree m, and, for ease of discussion, let's assume that its leading coefficient is 1. So f looks like this:

$$f(x) = x^m + a_{m-1}x^{m-1} + a_{m-2}x^{m-2} + \cdots + a_1 x + a_0.$$

By the fundamental theorem of algebra, f has m zeros (some may be the same and some may be complex numbers), say $r_1, \ldots, r_m$. By the factor theorem, it has m factors, $(x - r_1), \ldots, (x - r_m)$.

It follows that we have an identity:

$$\begin{aligned} x^m + a_{m-1}x^{m-1} + a_{m-2}x^{m-2} + \cdots + a_1 x + a_0 \\ = (x - r_1)(x - r_2)\cdots(x - r_m). \end{aligned} \tag{$*$}$$

In problem 33 on page 62, you multiplied the right-hand side of this identity out. Just for completeness, here are the details:

$$\begin{aligned} (x - r_1)(x - r_2)\cdots(x - r_m) = x^m& \\ &- r_1 x^{m-1} - r_2 x^{m-1} - r_3 x^{m-1} - \cdots - r_m x^{m-1} \\ &+ r_1 r_2 x^{m-2} + r_1 r_3 x^{m-2} + \cdots + r_1 r_m x^{m-2} + \cdots \\ &\qquad + r_{m-1} r_m x^{m-2} \\ &- r_1 r_2 r_3 x^{m-3} - r_1 r_2 r_4 x^{m-3} - \cdots \\ &\qquad - r_{m-2} r_{m-1} r_m x^{m-3} \\ &\vdots \quad \vdots \\ &\pm r_1 r_2 r_3 \cdots r_m. \end{aligned}$$

Ways to think about it

One way to think about multiplying all these binomials out is to think of all the ways you can take xs out of some parentheses and roots r_i out of the others. You can take

- all xs and no r_is (that produces the x^m on the top line),
- $m - 1$ xs and one r_is (that produces the second line),
- $m - 2$ xs and two r_is (that produces the third line),

$\vdots$

- no xs and all r_is (that produces the last line).

Now gather up like terms, writing this as a polynomial in x whose coefficients are expressions in the roots:

$$\begin{aligned} (x - r_1)(x - r_2)\cdots(x - r_m) = x^m& \\ &- (r_1 + r_2 + r_3 + \cdots + r_m)x^{m-1} \\ &+ (r_1 r_2 + r_1 r_3 + \cdots + r_1 r_m + \cdots + r_{m-1} r_m)x^{m-2} \\ &- (r_1 r_2 r_3 + r_1 r_2 r_4 + \cdots + r_{m-2} r_{m-1} r_m)x^{m-3} \\ &\vdots \quad \vdots \\ &+ (-1)^m r_1 r_2 r_3 \cdots r_m. \end{aligned}$$

There are two other common ways of writing this:

$$
\begin{aligned}
(x-r_1)(x-r_2)\cdots(x-r_m) = x^m &- \text{(the sum of the roots) } x^{m-1} \\
&+ \text{(the sum of the roots taken 2 at a time) } x^{m-2} \\
&- \text{(the sum of the roots taken 3 at a time) } x^{m-3} \\
&\vdots \quad \vdots \\
&+ (-1)^m \text{ (the sum of the roots taken } m \text{ at a time)}
\end{aligned}
$$

and

$$
\begin{aligned}
(x-r_1)(x-r_2)\cdots(x-r_m) = x^m &- \left(\sum_{i=1}^{m} r_i\right) x^{m-1} + \left(\sum_{i>j=1}^{m} r_i r_j\right) x^{m-2} \\
&- \left(\sum_{i>j>k=1}^{m} r_i r_j r_k\right) x^{m-3} \\
&\vdots \quad \vdots \\
&+ (-1)^m (r_1 r_2 r_3 \cdots r_m).
\end{aligned}
$$

Equating coefficients in (*), we have

Theorem 9. *If* $f(x) = x^m + a_{m-1}x^{m-1} + a_{m-2}x^{m-2} + \cdots + a_1 x + a_0$, *then*

$$
\begin{aligned}
a_{n-1} &= - \textit{ the sum of the zeros of } f, \\
a_{n-2} &= + \textit{ the sum of all products of two zeros of } f, \\
a_{n-3} &= - \textit{ the sum of all products of three zeros of } f, \\
\vdots \quad &\vdots \\
a_0 &= (-1)^m \textit{ times the product of the zeros of } f.
\end{aligned}
$$

So, the coefficients in a polynomial are functions of the zeros. In fact, they are special kinds of functions: sums, sums taken two at a time, sums taken three at a time, and so on. For a cubic, for example, if the roots are r, s, and t, these functions are

$$r+s+t$$

$$rs+rt+st$$

$$rst$$

These are called the *elementary symmetric functions* in r, s, and t. The elementary symmetric functions in r, s, t, and u are

$$r+s+t+u$$

$$rs+rt+ru+st+su+tu$$

$$rst + rsu + rtu + stu$$

$$rstu$$

These are called *symmetric* functions because if you permute the variables, you get the same expression back. There are other symmetric functions, like $r^2 + s^2$ or $r^3s + sr^3 - 5rs$. Our coefficient-generating functions are the *elementary* symmetric functions because of a fairly deep result:

We're assuming symmetric functions are *polynomials* in their variables. Actually, Theorem 10 is true in a more general context.

Theorem 10. *Every symmetric function can be expressed as a polynomial in the elementary symmetric functions that use the same variables.*

A proof of the symmetric function theorem is in [20].

So, for example,

$$r^2 + s^2 = (r + s)^2 - 2rs$$

and

$$\begin{aligned} r^3s + sr^3 - 5rs &= rs(r^2 + s^2 - 5) = rs\big((r + s)^2 - 2rs - 5\big) \\ &= rs(r + s)^2 - 2(rs)^2 - 5rs. \end{aligned}$$

Unfortunately, Theorem 10 doesn't tell you *how* to write a given symmetric expression in terms of the elementary symmetric functions—it just guarantees that there is a way to do it.

Ways to think about it

So, when you have an equation, there are numbers you don't know (the roots) and numbers you do know (the coefficients). And now we know how these two sets of numbers are related: the coefficients have "hidden meanings": they are the elementary symmetric functions of the roots. So, in a sense, solving an equation amounts to inverting this connection—untangling the symmetric functions to express the roots in terms of the coefficients. And that's what Abel showed couldn't be done for equations of degree 5 or more. *Very* roughly he showed that the roots of a generic equation were so unrelated to each other that there were many permutations of the roots that didn't affect the coefficients (120 of them for degree 5), and these permutations formed a system that couldn't be untangled—it couldn't be broken down into simpler pieces. And that prevented the existence of a "formula" involving only algebraic operations.

The term "hidden meaning" in this context is due to Paul Goldenberg.

Example: Let's find an equation whose roots are the squares of the roots of $x^3 - 5x^2 + 3x + 1 = 0$. Suppose the roots of this equation are r, s, and t. We could try to find the roots and square them, but that would be a lot of work. And we don't *need* each of the squares to find the equation satisfied by them—all we need is their elementary symmetric functions. That is, we want

$$r^2 + s^2 + t^2$$
$$r^2s^2 + r^2t^2 + s^2t^2$$
$$r^2s^2t^2$$

and what do we know? We know that

$$r + s + t = 5$$
$$rs + rt + st = 3$$
$$rst = -1$$

Ah, and we know (thanks to Theorem 10) that the three things we want can be expressed in terms of the three things we know. Finding out how to do it is where the fun is. Well, $r^2s^2t^2 = (rst)^2 = (-1)^2 = 1$. Let's do the sum of the squares. A good idea might be to start with the square of the sum:

$$(r + s + t)^2 = r^2 + s^2 + t^2 + 2rs + 2rt + 2st.$$

There we go:

$$r^2 + s^2 + t^2 = (r + s + t)^2 - 2(rs + rt + st) = 25 - 2 \cdot 3 = 19.$$

The "middle" coefficient is a little more stubborn, but let's follow our collective nose:

$$\begin{aligned}(rs + rt + st)^2 &= r^2s^2 + r^2t^2 + s^2t^2 + 2r^2st + 2rs^2t + 2rst^2 \\ &= r^2s^2 + r^2t^2 + s^2t^2 + 2rst(r + s + t).\end{aligned}$$

The fact that 19 showed up twice is pure accident. Or is it?

So,

$$\begin{aligned}r^2s^2 + r^2t^2 + s^2t^2 &= (rs + rt + st)^2 - 2rst(r + s + t) \\ &= 3^2 - 2 \cdot (-1) \cdot 5 = 19,\end{aligned}$$

and the equation we want is

$$x^3 - 19x^2 + 19x - 1.$$

Hmmm ... this almost looks like a reciprocal equation. It's not, but 1 is a root! We're in business. Hey, that means that 1 was a root of the original (because $1^2 = 1$). Why didn't we notice that?

It's not hard to see that these symmetric function calculations can get out of hand. Sometimes, there are other methods.

This may seem like a clever trick, but it actually generalizes to a method that lets you find the equation whose roots are any power of the roots of a given equation. See [14] for the details.

Example: Let's find an equation whose roots are the squares of the roots of $f(x) = x^4 - 5x^3 + x^2 - 3x - 4$. Symmetric functions would work, but the algebra would get hefty. Let's reason it out another way, using the identity $(*)$ on page 89. Let the zeros of f be r, s, t, and u. Then

$$f(x) = (x - r)(x - s)(x - t)(x - u)$$

and we *want* a function g so that

$$g(x) = (x - r^2)(x - s^2)(x - t^2)(x - u^2).$$

Look at the expression $f(x) \cdot f(-x)$:

$$f(x) \cdot f(-x) = ((x-r)(x-s)(x-t)(x-u))((-x-r)(-x-s)(-x-t)(-x-u)).$$

Factor a "–" out each of the last four binomials and regroup:

$$\begin{aligned} f(x) \cdot f(-x) &= ((x-r)(x-s)(x-t)(x-u))((x+r)(x+s)(x+t)(x+u)) \\ &= (x^2 - r^2)(x^2 - s^2)(x^2 - t^2)(x^2 - u^2). \end{aligned}$$

So, $f(x) \cdot f(-x)$ will have exactly the structure we want, but it will be a function of x^2, not x. Well, replace x^2 with x, and you'll get a function of x that factors exactly as we like. In our example,

A CAS will really help here.

$$f(x) \cdot f(-x) = x^8 - 23x^6 - 37x^4 - 17x^2 + 16,$$

so

$$g(x) = x^4 - 23x^3 - 37x^2 - 17x + 16$$

has zeros that are the squares of the zeros of f.

Problems

97. Suppose α, β, and γ are roots of the equation

Algebra texts from the 1920s and 1930s had pages and pages of problems like this.

$$x^3 - 4x^2 - 6x + 7 = 0.$$

Find the values of

(a) $\alpha^2 + \beta^2 + \gamma^2$

(b) $\alpha^2\beta^2 + \alpha^2\gamma^2 + \beta^2\gamma^2$

(c) $\alpha^3 + \beta^3 + \gamma^3$

(d) $\alpha^2\beta + \beta^2\alpha + \alpha^2\gamma + \gamma^2\alpha + \beta^2\gamma + \gamma^2\beta$

98. Show that

Why do we need to square everything?

$$((\alpha - \beta)(\alpha - \gamma)(\beta - \gamma))^2$$

is a symmetric function of α, β, and γ.

99. Find the elementary symmetric functions of the integers 0, 1, 2, 3, and 4.

Hint: Consider the polynomial $\binom{x}{5}$ from Chapter 1.

100. Show that the quadratic equation $ax^2 + bx + c = 0$ has equal roots if and only if $b^2 - 4ac = 0$.

101. Show that two roots of our "depressed" cubic

Hint: The sum of the roots is 0.

$$x^3 + px + q = 0$$

are equal if and only if $27q^2 + 4p^3 = 0$.

Problem 102 sets up a recurrence among sums of powers.

For example, if the roots of a cubic are α, β, and γ, its discriminant is

$$((\alpha-\beta)(\alpha-\gamma)(\beta-\gamma))^2$$

In algebra 2, you may *define* the discriminant as b^2-4c. Here, it's defined another way and you want to *prove* that it's given by this expression.

Problem 104 generalizes problem 62 on page 75. **Hint:** Express $f(x)$ as the product of the $x-\alpha_i$.

This is a repeat of problem 55 on page 75, just in case you didn't do it before. Now you could use problem 104.

As a test for seeing if roots are equal, the sign of the discriminant doesn't matter. And for degrees higher than 2, the famous "positive discriminant means real roots" that works for quadratics needs a lot of modification before it can be salvaged, so we won't worry about it.

102. Show that

$$\alpha^k+\beta^k=(\alpha+\beta)(\alpha^{k-1}+\beta^{k-1})-\alpha\beta(\alpha^{k-2}+\beta^{k-2}).$$

103. The *discriminant* of a polynomial f (denoted by $\delta(f)$) is the square of the product of the pairwise differences of its roots. Show that:

(a) the discriminant of a polynomial can be expressed in terms of its coefficients,

(b) the discriminant is 0 if and only if two zeros of the polynomial are equal,

(c) the discriminant of the polynomial x^2+bx+c is b^2-4c.

104. Suppose f is a polynomial of degree m with zeros $\alpha_1,\ldots,\alpha_m$. Use the product rule for the derivative (see page 73) to show that

$$f'(x)=\frac{f(x)}{x-\alpha_1}+\frac{f(x)}{x-\alpha_2}+\cdots+\frac{f(x)}{x-\alpha_m}.$$

105. Show that a polynomial has a multiple root if and only if it shares a common factor with its derivative.

106. Suppose f is a polynomial of degree m with zeros $\alpha_1,\ldots,\alpha_m$. Show that its discriminant is given by the formula

$$\delta(f)=f'(\alpha_1)\cdot f'(\alpha_2)\cdots f'(\alpha_m).$$

107. Show that the discriminant of the depressed cubic

$$x^3+px+q$$

is $27q^2+4p^3$.

108. Suppose f is a polynomial of degree m. Explain how $\delta(g)$ is related to $\delta(f)$ if

(a) g is obtained from f by scaling its roots by c,

(b) g is obtained from f by translating its roots by c.

109. What is the discriminant of $4x^3-5x^2+2x-1$?

Notes for Selected Problems

Notes for problem 4 on page 58. As a polynomial in y,

$$\begin{aligned}f(x,y)&=5x+10x^2+10x^3+5x^4+(5+20x+30x^2+20x^3+5x^4)y\\&\quad+(10+30x+30x^2+10x^3)y^2\\&\quad+(10+20x+10x^2)y^3+(5+5x)y^4.\end{aligned}$$

As a polynomial in x,

$$\begin{aligned}f(x,y)&=x^4(5+5y)+x^3(10+20y+10y^2)+x^2(10+30y+30y^2+10y^3)\\&\quad+x(5+20y+30y^2+20y^3+5y^4)+5y+10y^2+10y^3+5y^4.\end{aligned}$$

The factored form in 4a shows that the only x value that makes $f(x, y)$ vanish identically is $x = -1$. Same for y. Another way to see this is to look at the above expressions. In the first one, if f vanishes identically, it vanishes coefficient by coefficient. The only way you can get the coefficient of y^4 to be 0 is for x to be -1. Luckily, it makes all the other coefficients 0 too.

Notes for problem 5 on page 59. Suppose $f(x, y) = g(x, y)$. We write f and g as polynomials in y with coefficients that are polynomials in x:

$$f(x) = \sum_{j=0}^{n} f_j(x)y^j \quad \text{and} \quad g(x) = \sum_{j=0}^{n} g_j(x)y^j.$$

Then $f_j(x) = g_j(x)$ for all j. So upon replacing y^j with a_j, we obtain the desired result:

$$\sum_{j=0}^{n} f_j(x)a_j = \sum_{j=0}^{n} g_j(x)a_j.$$

Don't read too much into this result—it's easy to get fooled into misapplying it. For example, if $f(x, y) - (xy)^2$ and $g(x, y) = x^2y^2$. Then clearly $f = g$. Let $a_1 = 1$ and $a_2 = 2$:

$$f(x, y) = (xy)^2 \to (x \cdot 1)^2 = x^2,$$
$$g(x, y) = x^2y^2 \to x^2 \cdot 2 = 2x^2.$$

What went wrong?

Notes for problem 9 on page 59. Let f be a polynomial of degree n. If it has, say, $n+1$ factors, problem 6a tells us that the degree of f is at least $n+1$.

Notes for problem 12 on page 60. To show that these functions are equal, we must show that they produce the same output for each input in their domain. Here, the domain is the set of all functions.

So, let f be any function. Then, we have:

$$\begin{aligned}
((\Delta + \Delta^2) \circ (\Delta^3 + \Delta^4))(f) &= (\Delta + \Delta^2)((\Delta^3 + \Delta^4)(f)) \\
&= (\Delta + \Delta^2)(\Delta^3(f) + \Delta^4(f)) \\
&= \Delta(\Delta^3(f) + \Delta^4(f)) + \Delta^2(\Delta^3(f) + \Delta^4(f)) \\
&= (\Delta^4(f) + \Delta^5(f)) + (\Delta^5(f) + \Delta^6(f)) \\
&= \Delta^4(f) + 2\Delta^5(f) + \Delta^6(f) \\
&= (\Delta^4 + 2\Delta^5 + \Delta^6)(f).
\end{aligned}$$

Be sure you can provide the reason for each of the steps above.

Notes for problem 30 on page 62. First check that $g(0) = 0$ and $g(1) = 1$. To show that g produces the Fibonacci numbers, we need to show that $g(n+2) = g(n+1) + g(n)$. Let

$$\alpha = \frac{1+\sqrt{5}}{2} \quad \text{and} \quad \beta = \frac{1-\sqrt{5}}{2}$$

and verify that $\alpha^2 = \alpha + 1$ and $\beta^2 = \beta + 1$.

Then,

$$\begin{aligned} g(n+2) &= \frac{1}{\sqrt{5}}(\alpha^{n+2} - \beta^{n+2}) \\ &= \frac{1}{\sqrt{5}}(\alpha^n \alpha^2 - \beta^n \beta^2) \\ &= \frac{1}{\sqrt{5}}\big(\alpha^n(\alpha+1) - \beta^n(\beta+1)\big) \\ &= \frac{1}{\sqrt{5}}(\alpha^{n+1} - \beta^{n+1}) + \frac{1}{\sqrt{5}}(\alpha^n - \beta^n) \\ &= g(n+1) + g(n). \end{aligned}$$

Notes for problem 36 on page 67. Suppose $(x-a)$ is a factor of $f(x)$. Then $f(x) = (x-a)q(x)$ for some polynomial q. Thus, $f(a) = (a-a)q(a) = 0 \cdot q(a) = 0$.

Now, if $f(a) = 0$, then the remainder when $f(x)$ is divided by $(x-a)$ is 0. So, $(x-a)$ must be a factor of $f(x)$.

Notes for problem 41 on page 68. Think of f as a polynomial in C with coefficients that are polynomials in S. Dividing f by $C^2 + S^2 - 1$ (a "quadratic" in C) and using the remainder theorem, we have

$$f(S, C) = (C^2 + S^2 - 1)q(S, C) + a(S)C + b(S).$$

For the infinitely many pairs (S, C) so that, if $C^2 + S^2 = 1$, we have that

$$a(S)C + b(S) = 0.$$

Deduce that for infinitely many values of S between -1 and 1,

$$a(S)\sqrt{1-S^2} + b(S) = 0 \quad \text{and} \quad -a(S)\sqrt{1-S^2} + b(S) = 0.$$

Use this to show that $a(S)$ and $b(S)$ must each be identically 0.

Notes for problem 43 on page 68. The general proof is a bit messy, so let's try a proof that is general in principle. Letting

$$f(x) = a_3x^3 + a_2x^2 + a_1x + a_0$$

we have the following:

$$\begin{aligned} f(x) - f(a) &= a_3x^3 + a_2x^2 + a_1x + a_0 - (a_3a^3 + a_2a^2 + a_1a + a_0) \\ &= a_3(x^3 - a^3) + a_2(x^2 - a^2) + a_1(x - a) \\ &= a_3(x-a)(x^2 + ax + a^2) + a_2(x-a)(x+a) + a_1(x-a) \\ &= (x-a)\big[a_3(x^2 + ax + a^2) + a_2(x+a) + a_1\big] \end{aligned}$$

Thus, $g(x) = a_3(x^2 + ax + a^2) + a_2(x+a) + a_1$.

Notes for problem 46 on page 74. Let

$$f(x) = c_0 + c_1 x + c_2 x^2 + \cdots + c_m x^m.$$

We need to show that for any n $(0 \le n \le m)$,

$$c_n = \frac{f^{(n)}(0)}{n!}.$$

Differentiating $f(x)$ n times yields

$$\begin{aligned} f^{(n)}(x) &= n(n-1)\cdots 1 \cdot c_n + (n+1)(n)(n-1)\cdots 2 \cdot c_{n+1}x \\ &\quad + (n+2)(n+1)\cdots 3 \cdot c_{n+2}x^2 + \text{ (terms involving } x^3 \text{ and higher).} \end{aligned}$$

Note that all the terms before $c_n x^n$ vanish. Substituting $x = 0$, we obtain $f^{(n)}(0) = n(n-1)\cdots 1 \cdot c_n$ and so

$$c_n = \frac{f^{(n)}(0)}{n!}.$$

Notes for problem 58 on page 75. Let $h(x) = f(x) - g(x)$. Since $h(x)$ is a polynomial, and suppose

$$\begin{aligned} h(x) &= c_0 + c_1 x + c_2 x^2 + \cdots + c_m x^m, \\ h'(x) &= c_1 + 2c_2 x + \cdots + mc_m x^m. \end{aligned}$$

Since $h'(x) = f'(x) - g'(x) = 0$, we get $c_1 = c_2 = \cdots = c_m = 0$. So, $h(x) = c_0$, a constant.

Notes for problem 80 on page 82. Let m be the degree of f. We use Theorem 3, with the roles of x and a switched, to obtain

$$f(x+a) = f(x) + f'(x)a + \frac{f''(x)}{2!}a^2 + \cdots + \frac{f^{(m)}(x)}{m!}a^m.$$

Likewise, we have

$$\begin{aligned} f(x+a+1) &= f(x+(a+1)) \\ &= f(x) + f'(x)(a+1) + \frac{f''(x)}{2!}(a+1)^2 + \cdots \\ &\quad + \frac{f^{(m)}(x)}{m!}(a+1)^m. \end{aligned}$$

Subtract the two to obtain the desired result.

Notes for problem 84 on page 86. Let g be a polynomial whose roots are a_n times the roots of f. Then, according to Theorem 6, we can take $g(x)$ to be

$$\begin{aligned} g(x) &= a_n x^n + a_n a_{n-1} x^{n-1} + (a_n)^2 a_{n-2} x^{n-2} + \cdots + (a_n)^{n-1} a_1 x + (a_n)^n a_0 \\ &= a_n (x^n + a_{n-1} x^{n-1} + a_n a_{n-2} x^{n-2} + \cdots + (a_n)^{n-2} a_1 x + (a_n)^{n-1} a_0). \end{aligned}$$

Now let

$$h(x) = x^n + a_{n-1}x^{n-1} + a_n a_{n-2}x^{n-2} + \cdots + (a_n)^{n-2}a_1 x + (a_n)^{n-1}a_0.$$

Since $h(x)$ is a constant multiple of $g(x)$, it has the same roots as $g(x)$. Thus, h is a monic polynomial whose roots are a_n times the roots of f.

Notes for problem 86a on page 86. If $g(x)$ is the function obtained by decreasing the roots of f by k, we have

$$g(x) = f(k) + f'(k)x + \frac{f''(k)}{2}x^2.$$

So we want $f'(k) = 0$. Since $f'(x) = 2ax + b$, we have $k = -(b/2a)$. This gives us

$$\begin{aligned} g(x) &= \left[a\left(-\frac{b}{2a}\right)^2 + b\left(-\frac{b}{2a}\right) + c\right] + ax^2 \\ &= ax^2 + \frac{-b^2 + 4ac}{4a}. \end{aligned}$$

Let $h(x) = 4a \cdot g(x) = 4a^2x^2 + (-b^2 + 4ac)$, and note that h and g have the same roots because they are constant multiples of each other.

Notes for problem 86b on page 86. Using problem 84, we see that

$$x^2 + 4a^2(-b^2 + 4ac)$$

is a monic polynomial whose roots are $4a^2$ times that of $h(x)$.

Notes for problem 86c on page 86.

$$\begin{aligned} x^2 + 4a^2(-b^2 + 4ac) &= 0, \\ x^2 &= 4a^2(b^2 - 4ac), \\ x &= \pm 2a\sqrt{b^2 - 4ac}. \end{aligned}$$

Notes for problem 86d on page 86. Since the solutions in problem 86c are $4a^2$ times the roots of $h(x)$, the roots of $h(x)$ are

$$x = \pm\frac{2a\sqrt{b^2 - 4ac}}{4a^2} = \pm\frac{\sqrt{b^2 - 4ac}}{2a}.$$

Finally, we add $k = -(b/2a)$ to obtain the roots of f:

$$x = -\frac{b}{2a} \pm \frac{\sqrt{b^2 - 4ac}}{2a}.$$

Notes for problem 100 on page 93. One way to prove this is to use the quadratic formula. But here is another approach using the ideas from this section. First, rewrite the original equation as

$$x^2 + \frac{b}{a}x + \frac{c}{a} = 0.$$

If its roots are r and s, we have

$$r+s=-\frac{b}{a} \quad \text{and} \quad rs=\frac{c}{a}.$$

Assuming that $r=s$, $r+s=-(b/a)$ becomes $2r=-(b/a)$. Solving for r gives $r=-(b/2a)$. Squaring both sides yields $r^2=b^2/4a^2$.

Since $r=s$, $rs=c/a$ becomes $r^2=c/a$. Combining the two expressions for r^2 gives us

$$\frac{b^2}{4a^2}=\frac{c}{a}.$$

This simplifies to $b^2=4ac$ or $b^2-4ac=0$.

Starting with $b^2-4ac=0$ and working backwards will yield $r=s$. Try it.

Notes for problem 105 on page 94. From problem 104, we have

$$f'(x)=\frac{f(x)}{x-\alpha_1}+\frac{f(x)}{x-\alpha_2}+\cdots+\frac{f(x)}{x-\alpha_m}.$$

Suppose f and f' share a common factor. So, without loss of generality, suppose $(x-\alpha_1)$ is a factor of $f'(x)$. Therefore,

$$(x-\alpha_1) \text{ is a factor of } \left(\frac{f(x)}{x-\alpha_1}+\frac{f(x)}{x-\alpha_2}+\cdots+\frac{f(x)}{x-\alpha_m}\right).$$

Since $(x-\alpha_1)$ divides all other terms of the above sum, it must divide $f(x)/(x-\alpha_1)$ as well. Therefore, $(x-\alpha_1)^2$ divides $f(x)$. In other words, α_1 is a double root of f.

Now suppose f has a multiple root, say $x=\alpha_1$. Thus, it is divisible by $(x-\alpha_1)^2$. Therefore,

$$\frac{f(x)}{x-\alpha_1}, \frac{f(x)}{x-\alpha_2}, \ldots, \frac{f(x)}{x-\alpha_m}$$

are all divisible by $(x-\alpha_1)$. Thus, so is $f'(x)$.

3

Complex Numbers, Complex Maps, and Trigonometry

Introduction

Complex numbers—numbers of the form $a+bi$ where a and b are real numbers and $i^2 = -1$—emerged in mathematics as algebraic objects. First, they were useful in solving cubic equations. Later, it was found that every polynomial of degree n with real coefficients has exactly n roots in the complex numbers. This amazing fact—that the complex numbers are all you need to solve any polynomial with real coefficients—is known as the fundamental theorem of algebra. We won't prove the fundamental theorem in this chapter, because we want to go in a different direction. But you can find readable accounts in [2, 8].

"Exactly n roots" if you count multiple roots more than once. The sixth-degree equation $(x-3)^4(x+1)^2 = 0$ has six roots: 3, 3, 3, 3, -1, and -1.

Of course, you've now introduced a new kind of number into the mix. If you need complex numbers to solve polynomials with real coefficients, you probably need some other kind of number to solve polynomials with complex coefficients, right? Amazingly, no! Even if you allow the coefficients of a polynomial to be complex numbers, you still have exactly n roots for a polynomial of degree n, and those roots can all be found in the complex number system. Later in the chapter, we'll use the fundamental theorem of algebra to prove this generalized result.

In spite of its name, every proof of the fundamental theorem must use results that are strictly outside algebra.

Often in mathematics, the significance of new findings is not immediately clear. Complex numbers helped to solve many problems in algebra. But adding a geometric interpretation to these numbers allowed for new insights into other fields of study as well. In particular, the study of dynamics, chaos, iterated function systems, and fractals all involve functions on complex numbers.

Just as we wrote $f \in \mathbb{R}[x]$ in Chapter 2 to mean f is a polynomial in x with real coefficients, we'll write $f \in \mathbb{C}[x]$ to mean f is a polynomial in x with complex coefficients.

Complex numbers sit underneath many topics in school mathematics—from the very oldest topics in algebra to the newest visitors to school mathematics like fractals. Knowing your way around complex numbers helps you make sense of many topics and helps connect different ideas. In this chapter, we provide a short tour of complex numbers and some of their applications to algebra, geometry, analysis, and even arithmetic. We'll

The last bullet will involve an important family of polynomials, so we'll return to the themes of Chapter 2.

- look at them as algebraic tools for solving equations,
- find a geometric representation that connects to trigonometry and geometry,
- look at functions of complex numbers and some famous fractal curves, and
- use complex numbers to generate and prove trigonometric identities.

3.1 Complex Numbers

Complex numbers are numbers of the form $a+bi$ where a and b are real numbers and $i^2 = -1$. Most texts introduce the symbol i as a way to solve quadratic equations with no real roots. The first example is invariably

$$x^2 + 1 = 0.$$

This masks the history a bit. In fact, the "imaginary unit" i and complex numbers first came about in the solution to *cubic* equations.

Cardan's formula is to cubics what the quadratic formula is to quadratics: a way to find the roots for any cubic exactly, if the roots exist. Here's how it works: Suppose you have this cubic:

$$y^3 + by^2 + cy + d = 0.$$

A proof of Cardan's formula is developed in problems 88–90 on page 87 of Chapter 2.

A simple variable substitution (see problem 88 on page 87 of Chapter 2), $y = x - (b/3)$, allows you to get rid of the degree-2 term, so now you have

$$x^3 + px + q = 0.$$

Actually, there are *three* choices for each cube root; we're choosing one of them here.

Solutions to this equation are given by:

$$x = \sqrt[3]{\frac{-q + \sqrt{\frac{27q^2+4p^3}{27}}}{2}} + \sqrt[3]{\frac{-q - \sqrt{\frac{27q^2+4p^3}{27}}}{2}}.$$

Example: To solve $x^3 + x - 2 = 0$, note that $p = 1$ and $q = -2$. So Cardan's formula gives:

$$x = \sqrt[3]{\frac{2 + \sqrt{\frac{27(-2)^2+4}{27}}}{2}} + \sqrt[3]{\frac{2 - \sqrt{\frac{27(-2)^2+4}{27}}}{2}}.$$

Amazingly, this simplifies to 1! (Check it with a CAS.)

Since $x = 1$ is a root of the equation, $x - 1$ is a factor of the polynomial.

There are also two complex roots, which you can find by choosing different cube roots. Or, after finding that 1 is a root, you can divide the cubic by $x - 1$ to get

$$x^3 + x - 2 = (x - 1)(x^2 + x + 2) = 0.$$

You can then solve the remaining quadratic to find the other roots.

In applying Cardan's formula to $x^3 - 15x - 4 = 0$, Bombelli found this as a root:

$$x = \sqrt[3]{\frac{4 + \sqrt{\frac{27(-4)^2+4(-15)^3}{27}}}{2}} + \sqrt[3]{\frac{2 - \sqrt{\frac{27(-4)^2+4(-15)^3}{27}}}{2}}$$

$$= \sqrt[3]{2 + \sqrt{-121}} + \sqrt[3]{2 - \sqrt{-121}}.$$

Now, Bombelli knew that this polynomial had nice roots—in fact, 4 is one of the roots—so the square roots of negative numbers were puzzling. He was able to complete calculations like this by pretending that things like $\sqrt{-121}$ exist, but he disapproved of such methods.

Notice that, if we calculate formally,

$$\begin{aligned}(2 + \sqrt{-1})^3 &= 2^3 + 3 \cdot 2^2 \cdot \sqrt{-1} + 3 \cdot 2 \cdot (\sqrt{-1})^2 + (\sqrt{-1})^3 \\ &= 8 + 12\sqrt{-1} - 6 - \sqrt{-1} \\ &= 2 + 11\sqrt{-1} \\ &= 2 + \sqrt{-121}.\end{aligned}$$

In other words,

$$\sqrt[3]{2 + \sqrt{-121}} = 2 + \sqrt{-1}.$$

Well, $2 + \sqrt{-1}$ is *one* of the cube roots of $\sqrt[3]{2 + \sqrt{-121}}$.

Similarly,

$$\sqrt[3]{2 - \sqrt{-121}} = 2 - \sqrt{-1}.$$

If you make these substitutions into Cardan's solution, you get

$$x = (2 + \sqrt{-1}) + (2 - \sqrt{-1}) = 4.$$

As is always the case in mathematics, ideas created by one mathematician are refined and generalized by others. Euler was the first one to introduce the symbol i for $\sqrt{-1}$, and it took time for the algebra of the complex number system to be understood.

Just to get some perspective: Cardan lived from 1501–1575. The first rigorous proof of the Fundamental Theorem of Algebra—the fact that a polynomial of degree $n \geq 1$ (with real or complex coefficients) has exactly n roots in the complex numbers—was published by Gauss in 1799. It was more than 200 years after Bombelli worked with $\sqrt{-121}$ while computing the roots of a cubic.

Actually, Cardan should not be given much credit here. Scipio Ferro first discovered a general solution to cubics, but he preferred to keep it a secret shared by only his students. 40 years later, Tartaglia independently came up with the same method. He showed his method to Cardan who, despite his promise to secrecy, published the result.

Facts and Notation

We all agree that $\sqrt{9}$ means 3. So, the two square roots of 9 are $\sqrt{9}$ and $-\sqrt{9}$. This is just convention, part of the definition of $\sqrt{\quad}$.

Similarly, we agree that $\sqrt{-1} = i$, so that the two square roots of -1 are i and $-i$. And, one more convention: $\sqrt{-121} = 11i$, so the two square roots of

-121 are $11i$ and $-11i$. Similarly, $\sqrt{-3} = i\sqrt{3}$, and, if $c > 0$, $\sqrt{-c} = i\sqrt{c}$ (note: $\sqrt{c}$ is a real number). Just for typographical reasons, $\sqrt{-16}$ is written as $4i$, while $\sqrt{-17}$ is written as $i\sqrt{17}$.

Every complex number can be written as $a+bi$ with a and b real. A complex number written as $a+bi$ with a and b real is said to be in *standard form.*

If you know how to manipulate expressions like $3+4x$, you can add, subtract, and multiply complex numbers. Do the calculations as you usually would, and then whenever you have i^2, substitute -1.

$$(3+4i)+(2-i) = 5+3i$$
$$(3+4i)-(2-i) = 1+5i$$
$$(3+4i)(2-i) = 6+5i-4i^2 = 6+5i+4 = 10+5i$$

Proof If $a_1 + b_1 i = a_2 + b_2 i$ and $b_1 \neq b_2$, then i would equal

$$\frac{a_2 - a_1}{b_1 - b_2},$$

a nice real number. That's silly (why?). So, $b_1 = b_2$. This implies that $a_1 = a_2$ (why?).

If a complex number is written in standard form as $a+bi$ with a and b real numbers, a is called the *real part* and b is called the *imaginary part*. Two complex numbers $a_1 + b_1 i$ and $a_2 + b_2 i$ are equal if and only if $a_1 = a_2$ and $b_1 = b_2$. That is, for two complex numbers to be equal, the real parts must be equal and the imaginary parts must be equal.

If $z = a+bi$ is a complex number, then $\overline{z} = a-bi$ is known as the *conjugate* of z.

$$\begin{aligned} &\text{If } z = 2+3i, && \text{then } \overline{z} = 2-3i;\\ &\text{if } z = -1+\pi i, && \text{then } \overline{z} = -1-\pi i;\\ &\text{if } z = 2-7i, && \text{then } \overline{z} = 2+7i. \end{aligned}$$

Dividing complex numbers is not as straightforward as adding or multiplying. A number like

$$\frac{2-3i}{1+i}$$

is still a complex number, but to work with it, we want it to look like $a+bi$ with a and b real. Even if you separate the fraction above into two parts like this:

$$\frac{2}{1+i} - \frac{3i}{1+i},$$

you still don't know what is the "real part" and what is the "imaginary part."

What you want to do is get rid of the "imaginary part" in the denominator. You can do this by noticing a handy fact: When you multiply a complex number by its conjugate, you always get a real number. Check it out:

$$(2+3i)(2-3i) = 4-9i^2 = 4+9 = 13$$
$$(1-i)(1+i) = 1-i^2 = 1+1 = 2$$
$$(\sqrt{2}+\sqrt{3}i)(\sqrt{2}-\sqrt{3}i) = 3-3i^2 = 2+3 = 5$$

This is a lot like "rationalizing the denominator." Instead of getting rid of square root signs, we're getting rid of is.

So if you want to find

$$\frac{2-3i}{1+i},$$

you can multiply by $1 - i/1 - i$:

$$\frac{2-3i}{1+i} \cdot \frac{1-i}{1-i} = \frac{-1-5i}{2} = -\frac{1}{2} - \frac{5}{2}i.$$

This is in the $a + bi$ form, so it's a complex number we can deal with more easily.

Problems

1. Find one real root for each cubic using Cardan's formula.

(a) $x^3 - 18x - 35 = 0$ **(b)** $x^3 + 24x - 56 = 0$

2. Suppose $z = 1 - 4i$ and $w = 5 + 2i$. Do each calculation.

(a) $z + w$ **(b)** $z - w$ **(c)** zw

(d) z^2 **(e)** w^2 **(f)** $(z + w)^2$

3. **(a)** What is the conjugate of a real number? For example, if $z = -2$, what is $\overline{z}$? Explain your answer.

(b) What is the conjugate of a "pure imaginary" number? For example, if $z = 3i$, what is $\overline{z}$? Explain your answer.

(c) What is the conjugate of $\overline{z}$? Explain your answer.

(d) Characterize all complex numbers z so that $\overline{z} = z$.

(e) Find a complex number z so that $\overline{z} = z^2$.

4. Suppose $z = 2 - 3i$ and $w = 5 - i$. Do each calculation. Look for patterns.

(a) $z + \overline{z}$ **(b)** $z - \overline{z}$ **(c)** $z\overline{z}$

(d) $w + \overline{w}$ **(e)** $w - \overline{w}$ **(f)** $w\overline{w}$

(g) $\frac{z}{w}$ **(h)** $\frac{w}{z}$ **(i)** $\frac{z}{i}$

What is $\overline{i}$?

5. **(a)** State a conjecture about the product of a complex number and its conjugate: $z\overline{z}$.

(b) Prove your conjecture.

Hint: Let $z = a + bi$. What is $\overline{z}$? Now find the product and sum.

6. **(a)** State a conjecture about the sum of a complex number and its conjugate: $z + \overline{z}$.

(b) Prove your conjecture.

7. Prove the following theorem:

Theorem 1. *If z and w are complex numbers, then*

(a) $\overline{z + w} = \overline{z} + \overline{w}$, *and*

(b) $\overline{zw} = \overline{z}\,\overline{w}$

Say it like this: "The conjugate of the sum is the sum of the conjugates" and "the conjugate of the product is the product of the conjugates."

8. If $z = a + bi$ and $w = c + di$, express each in standard form (in terms of a, b, c, and d):

(a) $\frac{1}{z}$ **(b)** $\frac{w}{z}$

One way to go about this: Let $w = a + bi$. Find zw. Note that the real part must be 1 and the imaginary part must be 0. Another way: simplify $1/z$.

9. A **reciprocal** for z is a number w so that $zw = 1$. Try to find each reciprocal:

(a) $z = 1 + 2i$ **(b)** $z = 2 - i$ **(c)** $z = i$

10. Show that a quadratic equation with real coefficients and with one real root has, in fact, two real roots (that may possibly be the same).

Gaussian Integers are numbers of the form $a + bi$ where a and b are *integers*. For example, $27 - 15i$ is a Gaussian integer, but $3 + \frac{7}{2}i$ and $\sqrt{2} + i$ are not because $\frac{7}{2}$ and $\sqrt{2}$ are not integers. The following problems deal with Gaussian integers.

11. Is the sum of two Gaussian integers also a Gaussian integer? Explain how you know.

12. Is the product of two Gaussian integers also a Gaussian integer? Explain how you know.

13. Is the reciprocal of a Gaussian integer also a Gaussian integer? Explain how you know.

14. Note that in the Gaussian integers, you can factor some numbers that you usually think of as primes:

$$2 = (1 + i)(1 - i)$$

$$5 = (1 + 2i)(1 - 2i) = (2 + i)(2 - i)$$

(a) Which of these numbers are primes in the Gaussian integers and which are not: 3, 7, 11, 13, 17, 23, 29?

(b) Can you find a pattern? Which primes in the integers will still be primes in the Gaussian integers, and which will factor?

15. Pick five different Gaussian integers $a + bi$ where $a \neq b$. For each number you choose:

- Square your number to get a new number, $c + di$.
- Find $c^2 + d^2$. What kind of number is it?

16. Problem 15 suggests that squaring Gaussian integers is a way to generate Pythagorean triples. The conjecture is something like this: Suppose $a + bi$ is a Gaussian integer. If $(a + bi)^2 = c + di$, then $c^2 + d^2$ is a perfect square. In other words, c and d are the legs of some integer-sided right triangle.

(a) Prove this conjecture.

(b) Do you get all possible Pythagorean Triples this way?

Problem 16b is not so easy. See [12] for more details.

This is just the beginning. The Gaussian integers are just one example of algebraic systems that sit inside $\mathbb{C}$. These subsystems can be used to solve all kinds of problems from arithmetic and geometry. Reference [12] expands on the previous problem set and develops some general methods for generating other useful triples of numbers similar to Pythagorean triples.

Take it Further

On page 101 we mentioned the fact that the fundamental theorem of algebra guaranteed that every polynomial with *complex* coefficients had all its roots in $\mathbb{C}$. The next problem set helps you prove that if every polynomial with real coefficients has its roots in $\mathbb{C}$, so does every polynomial with complex coefficients.

17. Prove Lemma 1:

Lemma 1. *Suppose f is a polynomial with complex coefficients. Define $\overline{f}$ to be the polynomial you get by replacing each coefficient in f by its conjugate. Then:*

(a) $\overline{f+g} = \overline{f} + \overline{g}$.   *(b)* $\overline{fg} = \overline{f}\,\overline{g}$.
(c) $\overline{f} = f \Leftrightarrow f(x) \in \mathbb{R}[x]$.   *(d)* $f\overline{f} \in \mathbb{R}[x]$.
(e) $\overline{f(z)} = \overline{f}(\overline{z})$.

18. Assume that every polynomial with real coefficients and degree at least 1 has at least one root in $\mathbb{C}$. Use the Factor Theorem to show that a polynomial with real coefficients and degree n has n roots (if you count multiple roots as distinct) in $\mathbb{C}$.

Sometimes, the statement "Every polynomial with real coefficients and degree at least 1 has at least one root in $\mathbb{C}$" is taken *as* the statement of the fundamental theorem of algebra.

19. Suppose $f \in \mathbb{C}[x]$. If a complex number z is a root of f, show that $\overline{z}$ is a root of $\overline{f}$.

20. Suppose $f(x) \in \mathbb{C}[x]$. Then by Lemma 1, if we define $g(x) = f(x)\overline{f}(x)$, $g(x) \in \mathbb{R}[x]$. Show that if $g(z) = 0$, either $f(z) = 0$ or $\overline{f}(z) = 0$. Hence conclude that if every polynomial with real coefficients and degree at least 1 has a root in $\mathbb{C}$, then every polynomial with *complex* coefficients and degree at least 1 has a root in $\mathbb{C}$.

3.2 The Complex Plane

We want a way to represent complex numbers geometrically. Real numbers can be pictured as all the points along the number line.

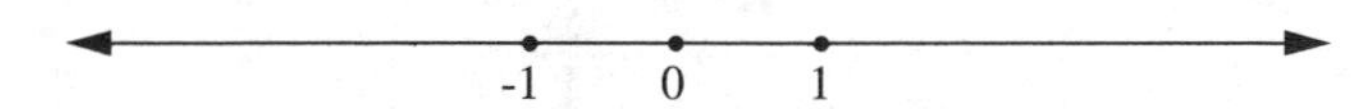

The absolute value of a number b, written $|b|$, is the distance of b from the point 0. So, $|b| = |-b|$.

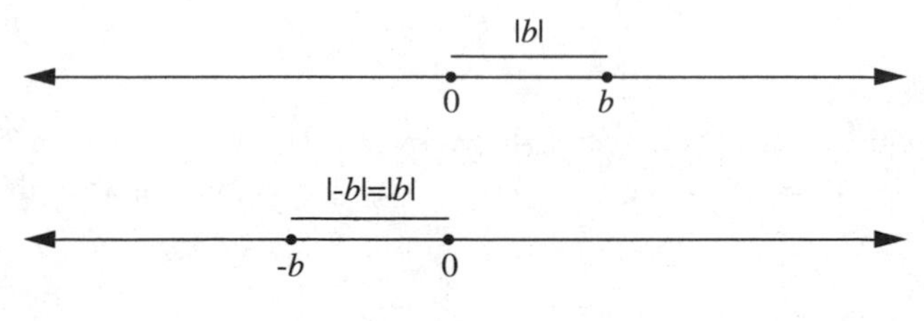

$|b|$ is the length of a segment

It took a couple of centuries after complex numbers were introduced into algebra before mathematicians (Gauss, Argand, and Wessel) hit upon this geometric representation. It may seem natural today, but it was quite a breakthrough in the history of the subject.

The designation of which is the real axis and which is imaginary is arbitrary. But it's important that everyone does the same thing.

To think about complex numbers, you need to think about *two* real numbers—the real and imaginary parts. One way to represent this situation is to use *two* number lines. The x-y plane below shows two real lines intersecting at the origin. The x-axis becomes the "real axis," where you graph the a (real part) of $a + bi$. The y-axis becomes the "imaginary axis," where you graph the b (imaginary part) of $a + bi$.

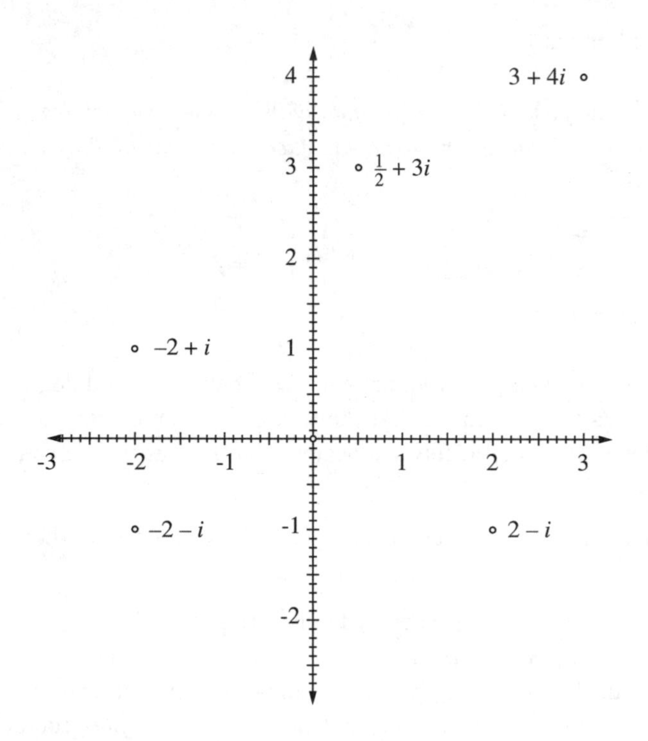

This representation is called the *complex plane*, and with it we can create an absolute value function for complex numbers. It measures the distance from a point $a + bi$ to the origin.

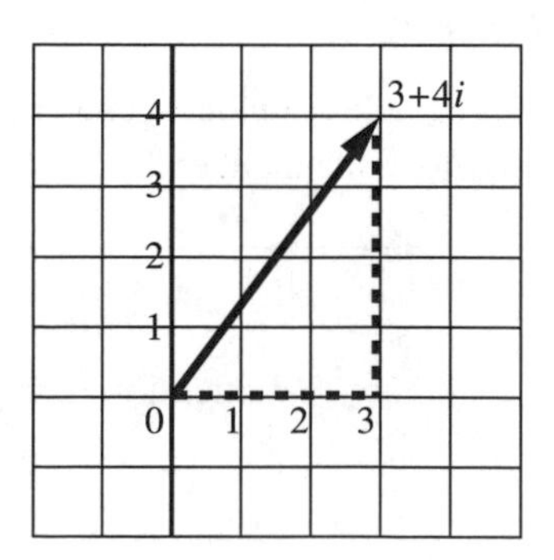

The symbol $|z|$ is usually read as the "absolute value of z" or the "length of z." In older books, it was sometimes called the "modulus of z."

Denote the length of the vector shown by $|z|$. By the Pythagorean Theorem, $|z|^2 = 3^2 + 4^2 = 25$, so $|z| = 5$. For a general complex number, this means $|z|^2 = a^2 + b^2$, so

$$|z| = \sqrt{a^2 + b^2}.$$

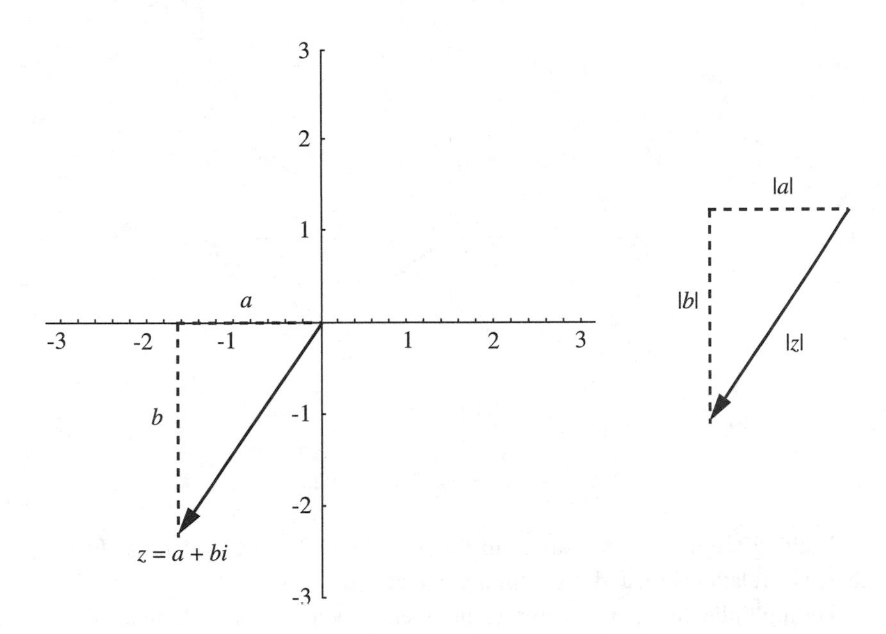

Now we have a picture of complex numbers as points in the plane. Notice that since every number $a + bi$ corresponds to a unique ordered pair (a, b) (and vice-versa), there is a one-to-one correspondence between points in the plane and complex numbers. Can we describe geometrically what addition and multiplication look like?

Just as there is a one-to-one correspondence between points on the number line and real numbers.

Well, if you add two complex numbers, you add the real parts and add the imaginary parts.

$$(a + bi) + (c + di) = (a + c) + (b + d)i.$$

This looks just like vector addition:

Proving this is the object of problem 21 on page 111.

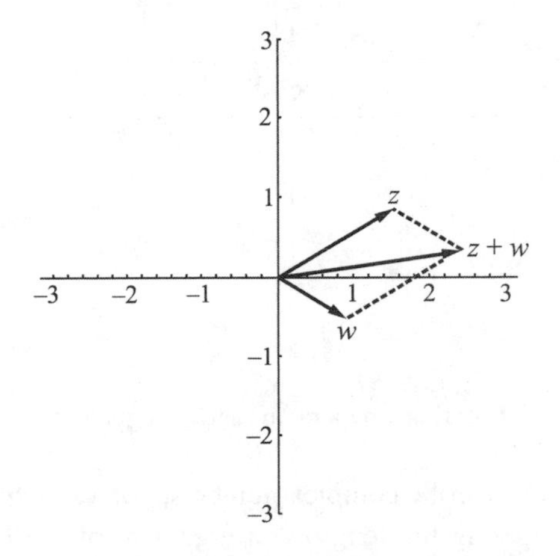

The vector $z + w$ is the diagonal of the parallelogram described by adjacent sides z and w.

Multiplication is trickier. Suppose $w = a + bi$ and $z = c + di$. We know that

$$wz = (a + bi)(c + di) = (ac - bd) + (ad + bc)i.$$

But is there a way to locate zw on the complex plane if you know z and w?

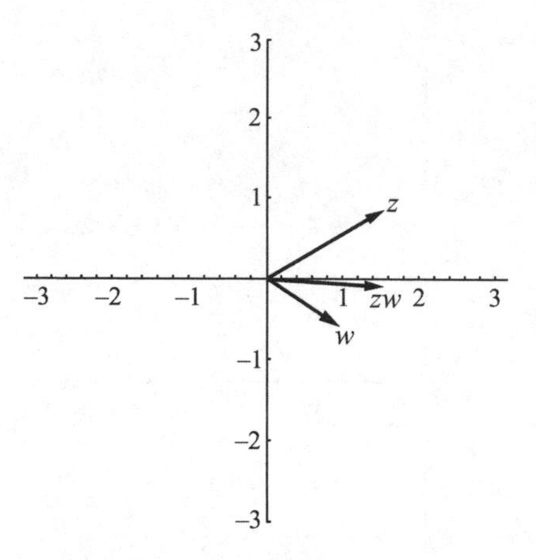

How do you get wz from w and z?

Nothing seems to jump out from this picture, or from the algebra. In fact, there is a relationship, and it's a simple one at that.

Notice that the radius of z is $|z|$. The argument is sometimes denoted by $\arg(z)$. So, $r = |z|$ and $\theta = \arg(z)$.

For multiplication of two complex numbers, it helps to think about characterizing the numbers in what's called **polar form**, rather than the standard $a + bi$ form. If z is a complex number, instead of saying how far over and up you go to get to z, you could describe how far z is from the origin (its *radius*) and what angle you have to turn through from the positive x-axis to get to z (it's *argument*). Hence, we can locate a complex number by an ordered pair (r, θ), where r is the length of the vector and θ is the angle it makes with the x-axis:

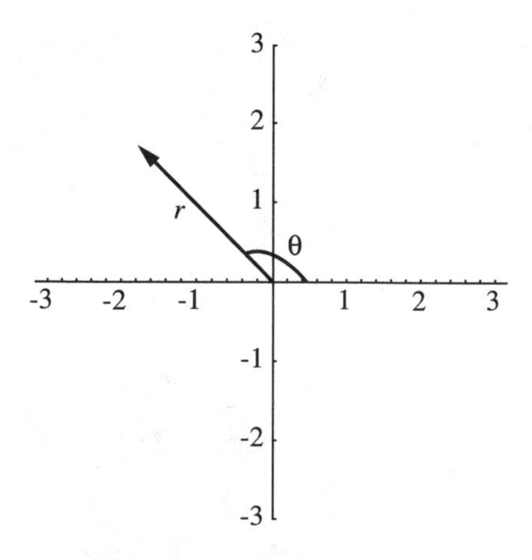

Locating z by a radius and an argument

Using this way to describe complex numbers, you can experiment with several examples, comparing the length and argument of zw to the lengths and arguments of z and w. What emerges is a pretty amazing conjecture. Let's stop here and work on some problems that bring together what we have and that investigate this conjecture.

Problems

21. Suppose $z = a + bi$ and $w = c + di$. Show that 0, z, w, and $z + w$ are the vertices of a parallelogram.

How can you show that a quadrilateral is a parallelogram?

22. For each number below, find

$$|z|, |\bar{z}|, |iz|, \text{ and } |-z|.$$

(a) $z = 5 + 12i$ **(b)** $z = -\sqrt{2} + \sqrt{2}i$ **(c)** $z = -\frac{7}{5} - \frac{24}{5}i$

23. Let $z = a + bi$ be any complex number. Show that $|z| = |\bar{z}| = |iz| = |-z|$.

24. Describe geometrically *all* the points with the same absolute value as z.

25. For each pair of numbers, find $|z|$, $|w|$, and $|zw|$.

(a) $z = 3 + 4i, \quad w = \sqrt{3} + i$ **(b)** $z = \sqrt{5} + 2i, \quad w = -i$

26. If $z \in \mathbb{C}$, show that $-z$ is symmetric to z with respect to the origin.

27. If $z \in \mathbb{C}$, show that $\bar{z}$ is symmetric to z with respect to the real axis.

28. Let $z = a + bi$ and $w = c + di$. Show that

(a) $|z||w| = |zw|$.

(b) Use this to show that $|z^n| = |z|^n$ for any nonnegative integer n.

29. Suppose z is a complex number such that $z^n = 1$ for some nonnegative integer n. Show that $|z| = 1$.

30. Let $z = a + bi$ and $w = c + di$. Use geometry to show that

(a) $|z + w| \le |z| + |w|$ **(b)** $|z + w| \ge |z| - |w|$

Under what conditions are these inequalities actual equalities?

The results of this problem are sometimes called the "triangle inequalities." Why?

31. For each number, find $|z|$ and $z\bar{z}$.

(a) $z = -3 - 4i$ **(b)** $z = \sqrt{3} - i$

(c) $z = 5$ **(d)** $z = -2i$

32. Let $z = a + bi$. Show that $|z| = \sqrt{z\bar{z}}$.

Problems 33–37 are experiments to help see the geometry behind the multiplication of complex numbers. Feel free to experiment on your own.

33. What happens to complex numbers on the circle of radius 5 (centered at the origin) when you multiply them by 2? How about when you multiply them by $\sqrt{2} - \sqrt{2}i$? As you would expect, multiplying by 2 is not the same as multiplying by $\sqrt{2} - \sqrt{2}i$. What differences do you notice?

34. **(a)** Choose five points on the circle of radius 1 and graph them.

(b) Multiply the points by 2 and graph the results. Describe how multiplying by 2 affects the location of the points.

(c) Multiply the points by $\sqrt{2} - \sqrt{2}i$ and graph the results. Describe how multiplying by $\sqrt{2} - \sqrt{2}i$ affects the location of the points.

35. Choose five points on the circle of radius 3, and repeat steps (b) and (c) of problem 34. Make a general conjecture about how multiplying by 2 and multiplying by $\sqrt{2} - \sqrt{2}i$ affect the locations of points in the complex plane.

You may want to choose some points on the circle of radius 1, as you did in problem 34, to form conjectures.

36. Describe (as completely as you can) how multiplying by z affects the locations of points in the complex plane.

(a) $z = \frac{1}{4}$ **(b)** $z = -1$ **(c)** $z = 3$ **(d)** $z = -5$

(e) $z = i$ **(f)** $z = -i$ **(g)** $z = -3i$ **(h)** $z = 1 + i$

(i) $z = -\frac{\sqrt{2}}{2} + \frac{\sqrt{2}}{2}i$ **(j)** $z = \sqrt{3} - i$

37. In general, describe how multiplying by a complex number $z = x + yi$ affects the locations of points in the complex plane. Be as specific as possible.

In these problems, $w = a + bi$ and $z = c + di$.

Problems 38–40 help you think about special cases of the problem of explaining the geometry behind the multiplication of complex numbers.

38. Suppose w is a real number, then $w = a + 0i$. So

$$wz = a(c + di) = (ac) + (ad)i.$$

Show that multiplying z by w "stretches" z a factor of $|a|$. (If $a < 0$, you have a stretch by $|a|$ and a rotation by 180°.)

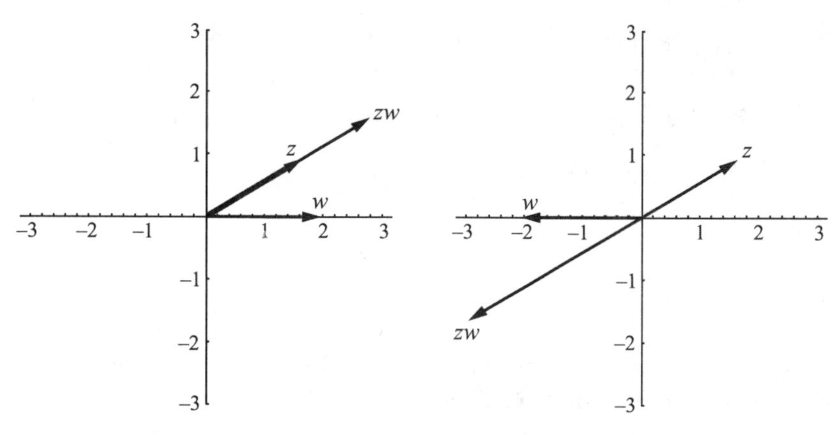

$w = a$. On the left, $a > 0$; on the right, $a < 0$.

In this picture, z is in the first quadrant. You should check the result for other quadrants.

39. Suppose next that $w = i$. Show that multiplication by i rotates 90° counterclockwise.

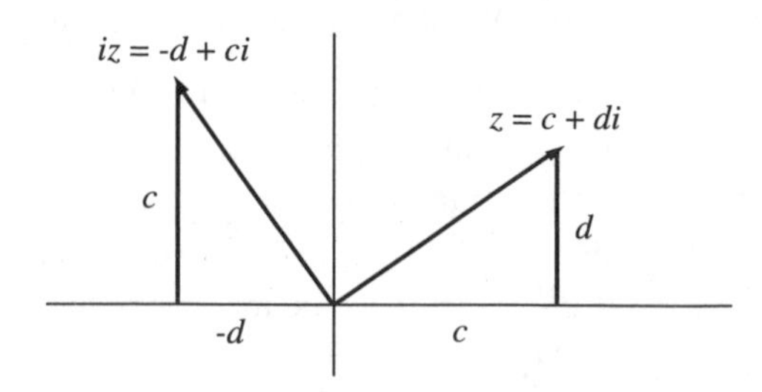

iz is obtained by rotating z 90° counterclockwise

Hint: Algebraically, $wz = i(c + di) = -d + ci$. Show that the slope of the vector from the origin to $c + di$ is the negative reciprocal of the slope of the vector from the origin to $-d + ci$, so they are perpendicular. A little more work (analyzing cases) shows which way wz points and that wz is obtained from z by a counterclockwise rotation of 90°.

40. More generally, if w is "pure imaginary" (that is $w = bi$), show that wz is obtained from z by rotating 90° counterclockwise and then stretching by a factor of $|b|$.

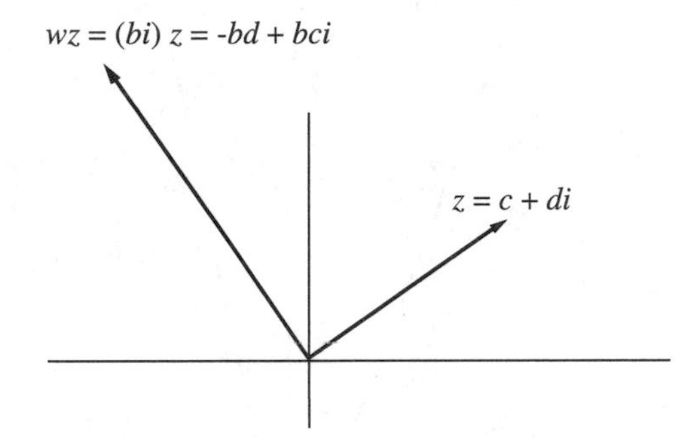

wz: rotate z 90° and stretch by $|b|$.

3.3 The Geometry behind Multiplying

It seems from the previous experiments that the length of zw is the product of the lengths of z and w and the argument of zw is the *sum* of the arguments of z and w. And, in fact, it's true. But getting there usually involves developing and using trig identities. In [15], we tried to circumvent the complications by getting at the essential mathematics beneath the complicated trig, and we succeeded in a way. But then, in the summer of 2002, a group of teachers at the Park City Mathematics Institute in Utah discovered a very simple and elegant way to see what's going on. Let's look at what they found.

Added in proof: The "PCMI" method does eppear elsewhere, for example in [17].

Suppose that $w = a + bi$ is some arbitrary complex number, and we want to see the effect of multiplying by w. The PCMI teachers reasoned like this:

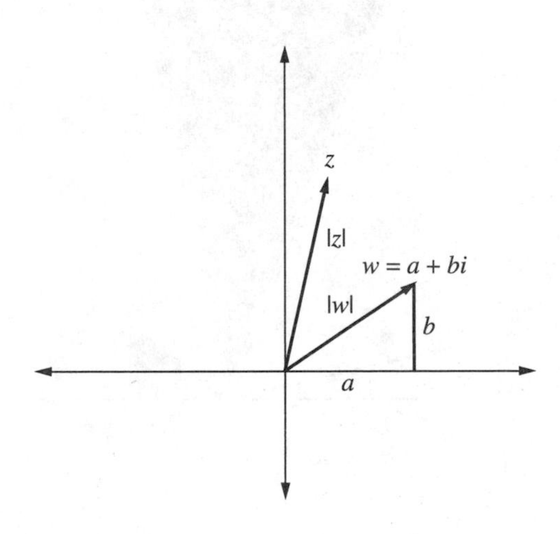

- Multiplying z by a scales it by a factor of $|a|$ (problem 38 from the previous section),
- Multiplying z by bi scales it by a factor of $|b|$ and rotates it 90° (problem 40 from the previous section), and
- $wz = (a + bi)z$ is the sum of az and $(bi)z$ (the distributive property in $\mathbb{C}$).

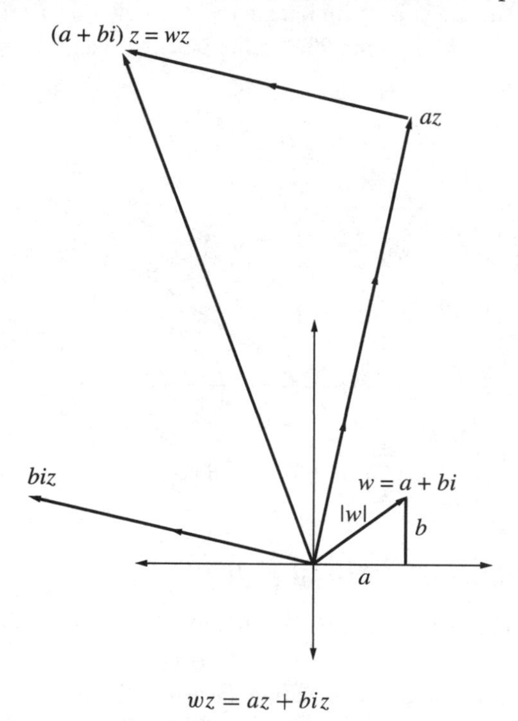

$$wz = az + biz$$

Now, the black triangles in the figure below are similar (SAS).

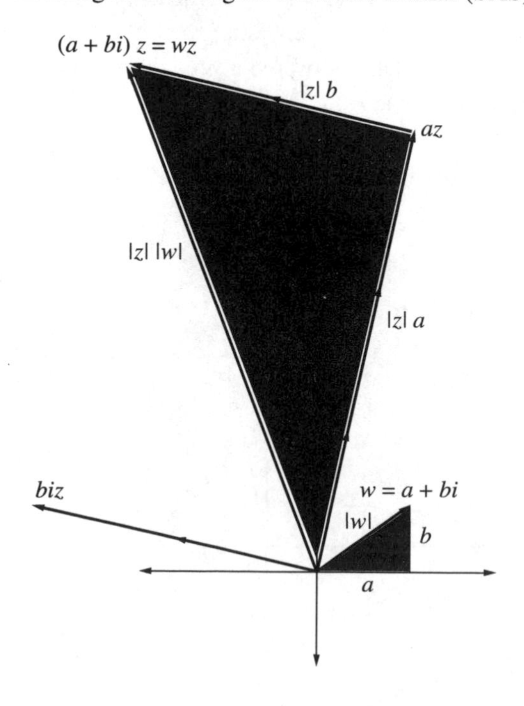

And the ratio of the sidelengths is $|z|$ (all the sides of the large one are $|z|$ times the corresponding sides of the small one). This implies that the hypotenuse of the large one has length $|z||w|$. But the length of this hypotenuse is just the absolute value of zw, so we have

$$|zw| = |z||w|.$$

An algebraic proof of this result is suggested in problem 28 on page 111.

As for the angles, notice that the argument of zw is the sum of the argument of z plus the black angle at the origin in the large triangle. And the black angle at the origin in the large triangle is the same as the black angle at the origin in the small triangle, because they are corresponding angles in similar triangles. But the black angle at the origin in the small triangle is nothing other than the argument of w. So, we have

$$\text{argument } (zw) = \text{argument } (z) + \text{argument } (w).$$

Remember, "the argument of a complex number" means the angle you get by rotating the real axis into the complex number in a counter-clockwise direction.

So, we have the following theorem:

Theorem 2. *If z and w are complex numbers, then the length of zw is the product of the lengths of z and w and the argument of zw is the sum of the arguments of z and w. In short,*

$$|zw| = |z||w| \quad \textit{and} \quad \arg(zw) = \arg(z) + \arg(w).$$

Some people say, "The length of the product is the product of the lengths, and the angle of the product is the sum of the angles."

Complex Numbers and Trigonometry

Theorem 2 can be used as a tool to simplify many results in trigonometry. To see how, let's review the connections between trigonometric functions and complex numbers.

More detailed treatments of these connections can be found in chapter 3 of [15] or in any good trigonometry book.

First, we'll focus just on complex numbers with absolute value 1; that is, the points on the unit circle. In the first quadrant, you have right triangles with

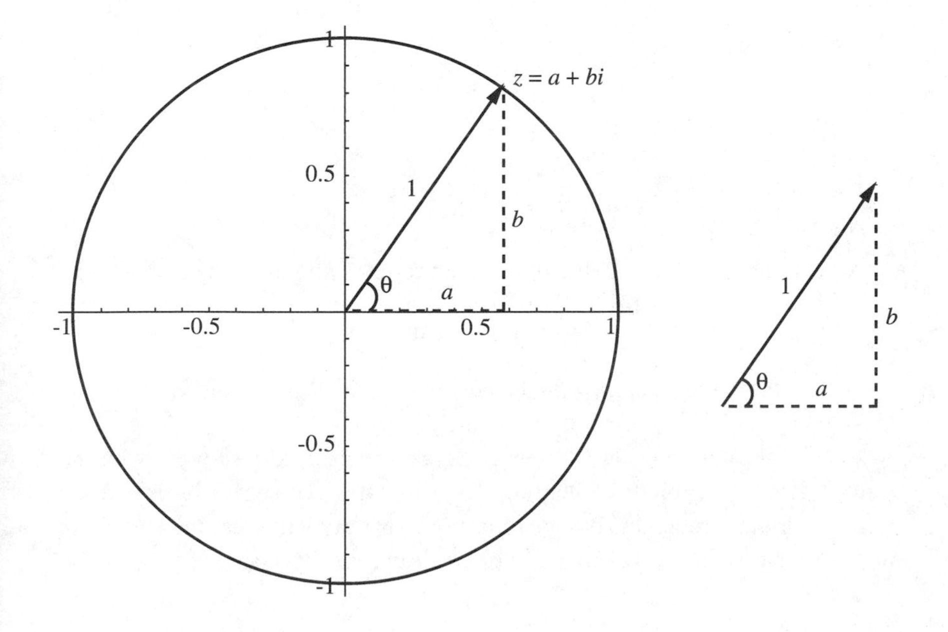

hypotenuse 1. So the lengths of the sides of the triangle (the a and b values) can be found directly from the angle at the origin:

$$a = \frac{a}{1} = \cos\theta = \frac{\text{adjacent}}{\text{hypotenuse}},$$

$$b = \frac{b}{1} = \sin\theta = \frac{\text{opposite}}{\text{hypotenuse}}.$$

We're assuming you know a bit of trig in what follows. This will not be a complete course, starting from scratch. Most of the details are in [15].

Of course, with these triangle-based definitions, sine and cosine don't make sense for angles greater than 90°. But we can *extend* the definitions of sine and cosine for these other angles based on the unit circle: Suppose $z = a + bi$ is a complex number on the unit circle, and z makes an angle of θ with the x-axis. We *define*:

$$\cos\theta = a: \text{ the } x\text{-coordinate of } z,$$

$$\sin\theta = b: \text{ the } y\text{-coordinate of } z.$$

So, we can write a and b in terms of the angle θ:

$$z = a + bi = \cos\theta + i\sin\theta.$$

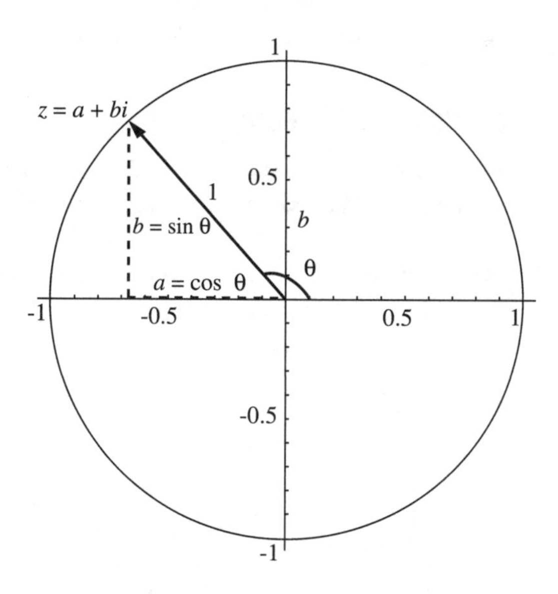

Now we have a trigonometric description for every point on the unit circle:

$$z = \cos\theta + i\sin\theta$$

where θ is the angle z makes with the positive x-axis. What about points not on the unit circle?

Each z corresponds to exactly one w on the unit circle.

Well, every point *not* on the unit circle can be mapped to a point with the same angle θ that *is* on the unit circle. To picture this geometrically, draw a ray from the origin through the point in question. That ray will cross the unit circle in exactly one point, and every point on that ray forms the same angle with the x-axis.

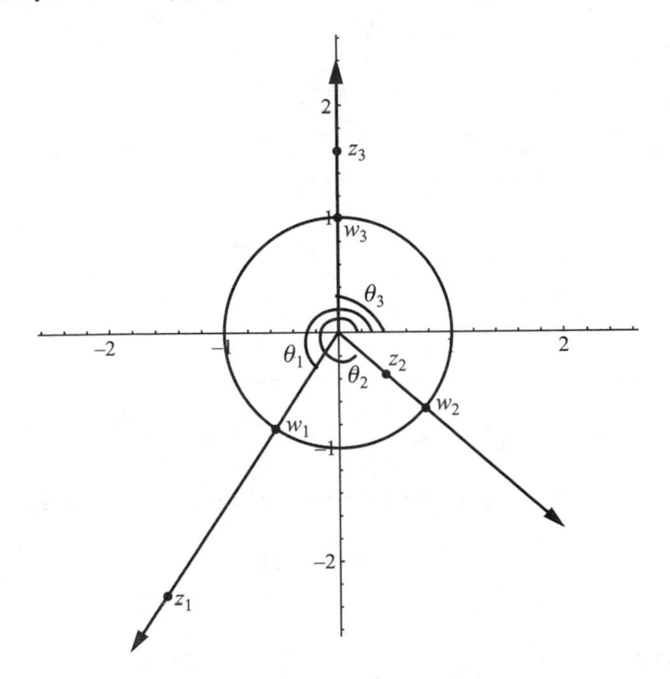

Algebraically, we can show that $w = z/|z|$ is on the unit circle. Suppose $z = a + bi$ is a nonzero complex number and $w = z/|z|$:

$$w = \frac{a+bi}{\sqrt{a^2+b^2}} = \frac{a}{\sqrt{a^2+b^2}} + \frac{b}{\sqrt{a^2+b^2}}i,$$

$$|w| = \sqrt{\left(\frac{a}{\sqrt{a^2+b^2}}\right)^2 + \left(\frac{b}{\sqrt{a^2+b^2}}\right)^2}$$

$$= \sqrt{\frac{a^2}{a^2+b^2} + \frac{b^2}{a^2+b^2}} = \sqrt{\frac{a^2+b^2}{a^2+b^2}} = 1.$$

Since $|w| = 1$, w is on the unit circle. So we can write

$$w = \cos\theta + i\sin\theta.$$

Since $z/|z| = w$, $z = |z|w$:

$$z = |z|w = |z|(\cos\theta + i\sin\theta).$$

Ways to think about it

There's another way to think about this result that saves some of the algebraic drudgery. It uses the *multiplicativity* of absolute value that was established in Theorem 2 (and in problem 28 on page 111). Let z be any complex number and let w be the complex number on the unit circle that corresponds to z. Then $w = cz$ for some real $c > 0$. Then we can argue like this:

(*continued*)

$$\begin{aligned} 1 &= |w| \\ &= |cz| = |c||z| \\ &= c|z|. \end{aligned}$$

So,

$$c = \frac{1}{|z|} \quad \text{and} \quad w = \frac{1}{|z|}z = \frac{z}{|z|}.$$

Summarizing, we have

Theorem 3. *Suppose z is a complex number and $\theta = \arg(z)$. Then*

$$z = |z|(\cos\theta + i\sin\theta).$$

Problems

Problems 41–43 are not just for "practice." There are patterns, and thinking about them gives you a feeling for the theorems.

Notice that in problems 41(l) and 41(m), we switched to radian measure. We'll gradually make the transition to using radians. Remember, radians measure distance around the unit circle, starting from $(1, 0)$, so that 2π radians is the same as 360°. For a refresher, see chapter 3 of [15].

41. Plot z, w, and zw, and then verify Theorem 2 for the following pairs z and w:

(a) $z = 1 + i,\ w = i$

(b) $z = 2 + 2i,\ w = i$

(c) $z = -2 + 2i,\ w = 1 + i$

(d) $z = 5i,\ w = -1 + i\sqrt{3}$

(e) $z = -4,\ w = \dfrac{-1 + i\sqrt{3}}{2}$

(f) $z = -2 - 2i,\ w = -1 - i$

(g) $z = \dfrac{-1 + i\sqrt{3}}{2},\ w = z$

(h) $z = \dfrac{-1 + i\sqrt{3}}{2},\ w = z^2$

(i) $z = 1 + i,\ w = \cos 30° + i\sin 30°$

(j) $z = \dfrac{-1 + i\sqrt{3}}{2},\ w = \bar{z}$

(k) $z = 2(\cos 30° + i\sin 30°),\ w = \cos 120° + i\sin 120°$

(l) $z = 2\left(\cos\dfrac{\pi}{4} + i\sin\dfrac{\pi}{4}\right),\ w = \cos\dfrac{3\pi}{2} + i\sin\dfrac{3\pi}{2}$

(m) $z = 2\left(\cos\dfrac{2\pi}{3} + i\sin\dfrac{2\pi}{3}\right),\ w = \cos\dfrac{\pi}{3} + i\sin\dfrac{\pi}{3}$

42. Verify Theorem 3 (and plot the numbers on the complex plane) for the following values of z:

(a) $1 + i$ **(b)** $2 + 2i$ **(c)** $2 - 2i$ **(d)** $5i$

(e) -3 **(f)** $-2 - 2i$ **(g)** $-1 + i\sqrt{3}$ **(h)** $\dfrac{-1 + i\sqrt{3}}{2}$

(i) $\dfrac{-1 - i\sqrt{3}}{2}$ **(j)** $3 + 4i$ **(k)** $3 - 4i$ **(l)** $-3 + 4i$

43. Calculate z, z^2, z^3, z^4, and z^5 and then plot each of these powers on the complex plane.

(a) $z = 1 + i$ **(b)** $z = \sqrt{2} + i\sqrt{2}$ **(c)** i

(d) $5i$

(e) -3

(f) $\dfrac{-1+i\sqrt{3}}{2}$

(g) $\dfrac{-1+i\sqrt{3}}{4}$

(h) $\dfrac{1+i\sqrt{3}}{2}$

(i) $\cos 20^\circ + i \sin 20^\circ$

(j) $\cos \dfrac{\pi}{6} + i \sin \dfrac{\pi}{6}$

(k) $\cos \dfrac{5\pi}{6} + i \sin \dfrac{5\pi}{6}$

(l) $\cos \dfrac{2\pi}{5} + i \sin \dfrac{2\pi}{5}$

44. Show that if $z = \cos\theta + i \sin\theta$ is a complex number on the unit circle,

$$\frac{1}{z} = \overline{z}.$$

Is the converse of this result true?

45. Suppose $z = \cos\theta + i\sin\theta$, and $w = \overline{z} = \cos\theta - i \sin\theta$.

(a) Show that z and w are roots of $x^2 - 2ax + 1$, where $a = \cos\theta$.

(b) Find (in terms of a), a quadratic equation whose roots are z^2 and w^2.

46. Prove Theorem 4:

Theorem 4. *For any value of* θ, $\sin^2\theta + \cos^2\theta = 1$.

The result of Theorem 4 is sometimes called the Pythagorean Identity. Why? **Hint**: Use problem 44.

47. Show that if $z = \cos\alpha + i\sin\alpha$ and $w = \cos\beta + i \sin\beta$ are complex numbers on the unit circle,

$$zw = \cos(\alpha+\beta) + i\sin(\alpha+\beta).$$

48. Prove Theorem 5:

Theorem 5. *For all values of* α *and* β,

$$\cos(\alpha+\beta) = \cos\alpha\cos\beta - \sin\alpha\sin\beta$$

and

$$\sin(\alpha+\beta) = \cos\alpha\sin\beta + \sin\beta\cos\alpha.$$

The results of Theorem 5 are called the *addition formulas for sine and cosine.* **Hint**: Use problem 47.

49. Prove Theorem 6 (*DeMoivre's theorem*):

Theorem 6. *If* n *is a positive integer, then*

$$(\cos\theta + i\sin\theta)^n = \cos n\theta + i \sin n\theta.$$

50. Use DeMoivre's theorem to solve the equation $x^6 - 1 = 0$.

51. Use DeMoivre's theorem to solve the equation $x^8 - 1 = 0$.

52. Use DeMoivre's theorem to solve the equation $x^8 - 128 = 0$.

For problems 49–52, plot the roots on the complex plane.

53. Show that the roots of $x^n - 1 = 0$ (n a positive integer) lie on the vertices of a regular n-gon on the complex plane.

54. If

$$z = \cos\frac{2\pi}{5} + i\sin\frac{2\pi}{5},$$

write z^{243} in the form $\cos\theta + i\sin\theta$.

3.4 Trigonometric Identities

And some don't . . .

Going back to the definitions uses the geometry of $\mathbb{C}$. Exploiting the multiplication rule uses its algebra.

What's the point to having students prove trigonometric identities? Sure, some are useful, and it's certainly true that some students get very good at the identity game. But isn't it just a bag of *ad-hoc* tricks?

There are two methods that bring some coherence (and even prettiness) to the topic of trigonometric identities. And the methods not only let students establish identities, they give them machines for generating their own. One is a special case of a bigger idea: *go back to the definitions*. Many teachers use a version of this technique, but it seems to be one of those "pick it up on the job" skills. The second seems to be fairly unknown in high school circles: *exploit the multiplication rule for complex numbers*. Let's first see how each works. Then we'll look at one special sequence of trig identities that shows up all over mathematics.

Go Back to the Definition

Remember that we extended definitions of sine and cosine from right triangles to the unit circle. $z = a+bi$ is a complex number on the unit circle, and z makes an angle of θ with the x-axis. We *define:*

$$\cos\theta = a: \text{ the } x\text{-coordinate of } z,$$

$$\sin\theta = b: \text{ the } y\text{-coordinate of } z.$$

So, we can write a and b in terms of the angle θ:

$$z = a + bi = \cos\theta + i\sin\theta.$$

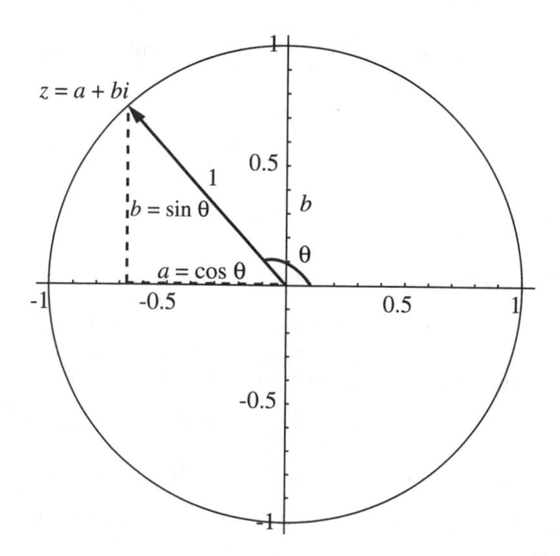

From this definition, we can form lots of trigonometric identities.

Example: Here is a picture that shows two ways to think about sine and cosine—on the unit circle, relating to the angle θ that a vector makes with the real axis, and in a right triangle, relating to an angle ψ that is less than $\pi/4$.

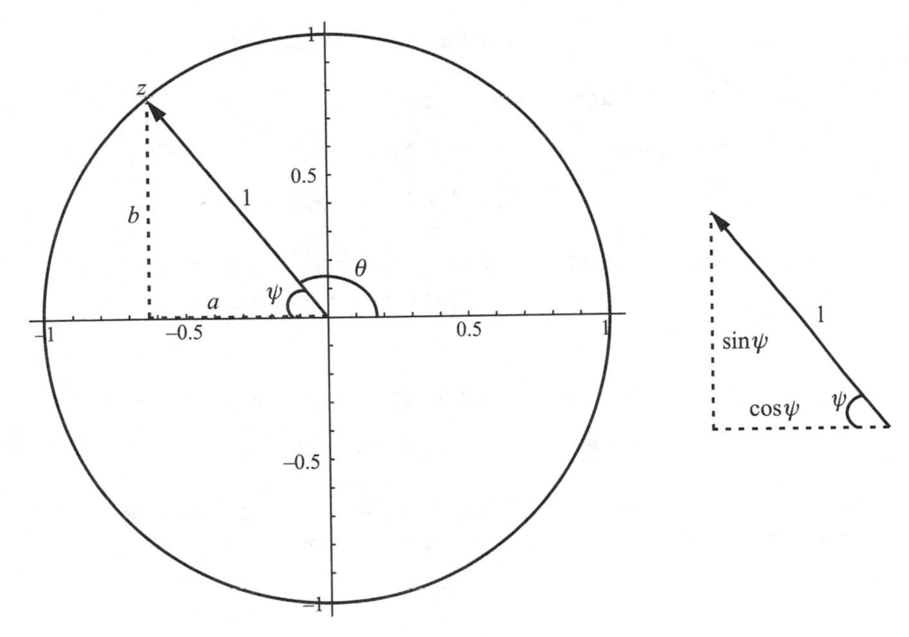

$a = \cos\theta$ is negative; $\cos\psi$ is positive.

In the picture, $\cos\theta = -\cos\psi$. (Since the point is in the second quadrant, a is negative, so $\cos\theta$ is negative. But for right triangles, sine and cosine are always between 0 and 1.) Also $\theta + \psi = \pi$. Putting these two things together, we get an identity:

So the lengths of the segments are the same, but the sign Is different.

$$\cos\theta = -\cos(\pi - \theta).$$

Similarly, $\sin\theta = \sin\psi$, since they are the same length and the same sign. In this case, we have:

$$\sin\theta = \sin(\pi - \theta).$$

So one way to prove identities is to draw pictures of the unit circle, pull out the appropriate right triangle, and look for relationships in the angles, the sines, and the cosines.

Exploit the Multiplication Rule for Complex Numbers

If we stick to the unit circle, suppose we have two numbers:

$$z = \cos\theta + i\sin\theta \qquad w = \cos\psi + i\sin\psi$$

Our multiplication rule says that to make zw, simply rotate z by an angle of ψ. (There is no stretching because $|w| = 1$.)

$$zw = \cos(\theta + \psi) + i\sin(\theta + \psi)$$

If you did problem 48 on page 119, this will look very familiar.

But suppose we just perform the multiplication, algebra style:

$$\begin{aligned} zw &= (\cos\theta + i\sin\theta)(\cos\psi + i\sin\psi) \\ &= \cos\theta\cos\psi + i\cos\theta\sin\psi + i\sin\theta\cos\psi + i^2\sin\theta\sin\psi \\ &= (\cos\theta\cos\psi - \sin\theta\sin\psi) + i(\cos\theta\sin\psi + \sin\theta\cos\psi) \end{aligned}$$

If both of these expressions are equal to zw, then the real parts must be equal and the imaginary parts must be equal, giving us the celebrated *addition formulas for sine and cosine*:

$$\begin{aligned} \cos(\theta+\psi) &= \cos\theta\cos\psi - \sin\theta\sin\psi \\ \sin(\theta+\psi) &= \cos\theta\sin\psi + \sin\theta\cos\psi \end{aligned}$$

This is just the beginning. For example, here's one way to see that cosine is even and sine is odd. Let $z = \cos\theta + i\sin\theta$. Because z is on the unit circle,

$$\frac{1}{z} = \bar{z}.$$

Notice:

$$\begin{aligned} 1 &= (\cos\theta + i\sin\theta) \\ &\quad \times (\cos\theta - i\sin\theta) \\ &= \cos^2\theta + \sin^2\theta. \end{aligned}$$

So, $1 = z\bar{z}$, and the conjugate of z is its reciprocal. But by the multiplication rule,

$$(\cos\theta + i\sin\theta)\big(\cos(-\theta) + i\sin(-\theta)\big) = \cos 0 + i\sin 0 = 1.$$

So, $\cos(-\theta) + i\sin(-\theta)$ is also the reciprocal of z. But z has only one reciprocal, so

$$\cos(-\theta) + i\sin(-\theta) = \cos\theta - i\sin\theta.$$

Comparing real and imaginary parts, we have:

$$\begin{aligned} \cos(-\theta) &= \cos\theta \qquad \text{(cosine is an } \textit{even} \text{ function)} \\ \sin(-\theta) &= -\sin\theta \qquad \text{(sine is an } \textit{odd} \text{ function)} \end{aligned}$$

Ways to think about it

Notice how we are standing the development on its head. Most developments of the multiplication rule *use* the addition formulas. We used Theorem 2 on page 115 instead. So we're free to use the multiplication rule for complex numbers to prove the addition formulas. And this exhibits a common device in mathematics. Just as Cardan and Bombelli could solve cubics by "moving up" to the complex numbers and then moving back down again, our method establishes equality of functions defined in $\mathbb{R}$—the trigonometric functions—by doing calculations in $\mathbb{C}$.

Rumor has it that all trigonometric identities can be reduced to complex number calculations, and that this is the basis for *Mathematica*'s trig-identity checker.

Example: Suppose you want to *find* the "double angle formulas," for cosine and sine; that is, formulas for $\cos 2\theta$ and $\sin 2\theta$ in terms of $\sin\theta$ and $\cos\theta$. You could replace ψ by θ in the addition formulas, but one calculation with complex numbers will give you both formulas at once: Just multiply $\cos\theta + i\sin\theta$ by itself:

$$\begin{aligned}\cos 2\theta + i\sin 2\theta &= (\cos\theta + i\sin\theta)(\cos\theta + i\sin\theta)\\ &= (\cos^2\theta - \sin^2\theta) + i(\sin\theta\cos\theta + \cos\theta\sin\theta)\\ &= (\cos^2\theta - \sin^2\theta) + 2i\sin\theta\cos\theta.\end{aligned}$$

So,

$$\cos 2\theta = \cos^2\theta - \sin^2\theta$$

and

$$\sin 2\theta = 2\sin\theta\cos\theta.$$

The formula for $\cos 2\theta$ is often re-written:

$$\cos 2\theta = \cos^2\theta - \sin^2 = \cos^2\theta - (1 - \cos^2\theta) = 2\cos^2\theta - 1.$$

Multiple Angle Formulas

For extensions of this idea, see [14].

What about $\cos 3\theta$? Or, more generally, $\cos n\theta$? Or $\sin n\theta$? It turns out that we can get these "trig identities" by combining some ideas from Chapter 2 with DeMoivre's theorem.

Let's start with $\cos 3\theta$. We can try to proceed inductively, expanding $\cos(3\theta)$ as $\cos(\theta + 2\theta)$:

$$\begin{aligned}\cos(3\theta) &= \cos(\theta + 2\theta)\\ &= \cos\theta\cos(2\theta) - \sin\theta\sin(2\theta)\\ &= \cos\theta(2\cos^2\theta - 1) - \sin\theta(2\sin\theta\cos\theta)\\ &= 2\cos^3\theta - \cos\theta - 2\sin^2\theta\cos\theta\\ &= 2\cos^3\theta - \cos\theta - 2(1 - \cos^2\theta)\cos\theta\\ &= 4\cos^3\theta - 3\cos\theta.\end{aligned}$$

So, $\cos(3\theta)$ is a polynomial (a cubic, in fact) in $\cos\theta$. Oh my. We have

$\cos 1\theta$	$\cos\theta$
$\cos 2\theta$	$-1 + 2(\cos\theta)^2$
$\cos 3\theta$	$-3(\cos\theta) + 4(\cos\theta)^3$

Will $\cos 4\theta$ be a fourth degree polynomial in $\cos\theta$?

We could keep going, but already the algebra is getting hefty. Before we invest in more hand calculation, let's use the CAS to generate some data. The question on the table is "Can $\cos(n\theta)$ be expressed as a polynomial (maybe of degree n) in $\cos\theta$?"

In *Mathematica*, there are functions like `TrigExpand` and `TrigSimplify`; on the TI machines, it's `tExpand` and `tSimplify`.

Most CAS implementations have functions that tell the algebraic manipulation routines that trigonometric functions should be treated as rational functions of sine and cosine.

Here's how they work (on the TI-89):

$$\texttt{tExpand (cos (}3\theta\texttt{))}$$

returns

$$\texttt{cos (}\theta\texttt{)} - 4(\texttt{sin(}\theta\texttt{)})^2 \cdot \texttt{cos(}\theta\texttt{)}$$

This isn't exactly what we got by hand, but it can be turned into it by replacing $\sin^2\theta$ by $1 - \cos^2\theta$. Let's try it for $\cos 4\theta$. Typing

$$\texttt{tExpand cos (}4\theta\texttt{)}$$

produces

$$1 - 8(\texttt{sin(}\theta\texttt{)})^2 \cdot (\texttt{cos(}\theta\texttt{)})^2$$

Even before we simplify it, we can see that this will be a polynomial of degree 4 in $\cos\theta$. It looks as if we have a conjecture in the wings. Here's what we have right now (after simplifying the above expression):

$\cos 1\theta$	$\cos\theta$
$\cos 2\theta$	$-1 + 2(\cos\theta)^2$
$\cos 3\theta$	$-3(\cos\theta) + 4(\cos\theta)^3$
$\cos 4\theta$	$1 - 8(\cos\theta)^2 + 8(\cos\theta)^4$

Form versus function again.

What we care about is the *form* of the (alleged) polynomials. What are their degrees? What can we say about the coefficients? For questions like this, "$\cos\theta$" is just a distraction; it might as well be an x or a y or any other symbol. Generating a few more of these formulas with a mix of hand and CAS calculations, and replacing $\cos\theta$ by x, we get

1	x
2	$-1 + 2x^2$
3	$-3x + 4x^3$
4	$1 - 8x^2 + 8x^4$
5	$5x - 20x^3 + 16x^5$
6	$-1 + 18x^2 - 48x^4 + 32x^6$
7	$-7x + 56x^3 - 112x^5 + 64x^7$

Now we have something to play with. Let's look for a recurrence. Looking at the highest degree terms:

$$x, 2x^2, 4x^3, 8x^4, \dots,$$

maybe we can get from one polynomial to the next by multiplying by $2x$. To make things easier, let's call the nth polynomial t_n. Then we can organize our data like this:

n	$t_n(x)$	$2xt_{n-1}(x)$
1	x	
2	$-1+2x^2$	$2x^2$
3	$-3x+4x^3$	$-2x+4x^3$
4	$1-8x^2+8x^4$	$-6x^2+8x^3$
5	$5x-20x^3+16x^5$	$2x-16x^3+16x^5$
6	$-1+18x^2-48x^4+32x^6$	$10x-40x^4+32x^6$
7	$-7x+56x^3-112x^5+64x^7$	$-2x+32x^3-96x^5+64x^6$

Well, this gives the right leading term, but the rest are not right, so we have a "first approximation." A good habit is to see "how far off" we are by subtracting the approximation from the real thing. Here we go:

n	$t_n(x)$	$2xt_{n-1}(x)$	$2xt_{n-1}(x)-t_n(x)$
1	x		
2	$-1+2x^2$	$2x^2$	1
3	$-3x+4x^3$	$-2x+4x^3$	x
4	$1-8x^2+8x^4$	$-6x^2+8x^3$	$-1+2x^2$
5	$5x-20x^3+16x^5$	$2x-16x^3+16x^5$	$-3x+4x^3$
6	$-1+18x^2-48x^4+32x^6$	$10x-40x^4+32x^6$	$1-8x^2+8x^4$
7	$-7x+56x^3-112x^5+64x^7$	$-2x+32x^3-96x^5+64x^6$	$5x-20x^3+16x^5$

Bingo. After t_2 at least, we have what seems to be a *two*-term recurrence

$$t_n(x) = 2xt_{n-1}(x) - t_{n-2}(x).$$

So, we have a sequence of polynomials:

$$t_k(x) = \begin{cases} x & \text{if } k = 1 \\ 2x^2 - 1 & \text{if } k = 2 \\ 2xt_{k-1}(x) - t_{k-2}(x) & \text{if } k > 2 \end{cases}$$

and we'd *very* much like it if

$$t_k(\cos\theta) = \cos k\theta \quad \text{for positive integers } k.$$

Well, we know this holds if k is any integer from 1 to 7. Now what? Well, let's get an expression for $\cos k\theta$ that might lend itself to a recurrence on k that's similar (the same, maybe) as that on the t_k.

Suppose we let $\alpha = \cos\theta + i\sin\theta$. Then by DeMoivre,

$$\alpha^k = \cos k\theta + i\sin k\theta.$$

We want just $\cos k\theta$. Since $\sin(-\theta) = -\sin\theta$, we know that if $\beta = \overline{\alpha} = \cos\theta - i\sin\theta$, then

$$\beta^k = \cos k\theta - i\sin k\theta.$$

Things are looking up.

So,

$$\alpha^k + \beta^k = 2\cos k\theta$$

which is twice what we want. And, since we want to relate this to $\cos(k-1)\theta$ and $\cos(k-2)\theta$, we'll want to look at two other instances of this:

$$\alpha^{k-1} + \beta^{k-1} = 2\cos(k-1)\theta \quad \text{and} \quad \alpha^{k-2} + \beta^{k-2} = 2\cos(k-2)\theta.$$

But these sums can be related by algebra. Indeed, in problem 102 on page 94 of Chapter 2, you showed that (no matter what α and β are)

$$\alpha^k + \beta^k = (\alpha+\beta)(\alpha^{k-1} + \beta^{k-1}) - \alpha\beta(\alpha^{k-2} + \beta^{k-2}).$$

But, for our $\alpha = \cos\theta + i\sin\theta$ and $\beta = \overline{\alpha}$, we have

$$\alpha + \beta = 2\cos\theta \quad \text{and} \quad \alpha\beta = 1.$$

Making all the substitutions, we find

$$2\cos k\theta = 2\cos\theta\big(2\cos(k-1)\theta\big) - \big(2\cos(k-2)\theta\big)$$

or

$$\cos k\theta = 2\cos\theta \cdot \cos(k-1)\theta - \cos(k-2)\theta.$$

Once we nail down the base cases, we have a nice recurrence for $\cos k\theta$:

$$\cos k\theta = \begin{cases} \cos\theta & \text{if } k = 1, \\ 2\cos^2\theta - 1 & \text{if } k = 2, \\ 2\cos\theta \cdot \cos(k-1)\theta - \cos(k-2)\theta & \text{if } k > 2. \end{cases}$$

How nice. This is exactly the same recurrence as satisfied by our t_k:

$$t_k(x) = \begin{cases} x & \text{if } k = 1, \\ 2x^2 - 1 & \text{if } k = 2, \\ 2xt_{k-1}(x) - t_{k-2}(x) & \text{if } k > 2. \end{cases}$$

So, by induction,

Remember, we verified this for k between 1 and 7.

$$t_k(\cos\theta) = \cos k\theta \quad \text{for } k > 0.$$

In fact, if we take $t_0(x)$ to be 1, this statement is even true for $k = 0$. So, we have a beautiful theorem:

Theorem 7. *The sequence of polynomials defined by*

$$t_k(x) = \begin{cases} 1 & \text{if } k = 0 \\ x & \text{if } k = 1 \\ 2xt_{k-1}(a) - t_{k-2}(x) & \text{if } k > 2 \end{cases}$$

has the property that

$$t_k(\cos\theta) = \cos k\theta.$$

The t_k are the *Chebyshev polynomials*. Much is known about these polynomials; we've seen that they can be used as a machine for generating the double, triple, quadruple, ... angle formulas for cosine:

For a biography of Chebyshev (and for a really useful website), visit `http://www-groups.dcs.st-and.ac.uk/history/`

$$\cos 2\theta = t_2(\cos\theta) = 2\cos^2\theta - 1$$

$$\cos 3\theta = t_3(\cos\theta) = 4\cos^3\theta - 3\cos\theta$$

$$\cos 4\theta = t_3(\cos\theta) = 8\cos^4\theta - 8\cos^2\theta + 1$$

$$\vdots \quad \vdots$$

Just for fun, here's a list of some multiple angle formulas, generated with a CAS:

$\cos 2\theta$	$-1 + 2(\cos\theta)^2$
$\cos 3\theta$	$-3(\cos\theta) + 4(\cos\theta)^3$
$\cos 4\theta$	$1 - 8(\cos\theta)^2 + 8(\cos\theta)^4$
$\cos 5\theta$	$5(\cos\theta) - 20(\cos\theta)^3 + 16(\cos\theta)^5$
$\cos 6\theta$	$-1 + 18(\cos\theta)^2 - 48(\cos\theta)^4 + 32(\cos\theta)^6$
$\cos 7\theta$	$-7(\cos\theta) + 56(\cos\theta)^3 - 112(\cos\theta)^5 + 64(\cos\theta)^7$
$\cos 8\theta$	$1 - 32(\cos\theta)^2 + 160(\cos\theta)^4 - 256(\cos\theta)^6 + 128(\cos\theta)^8$
$\cos 9\theta$	$9(\cos\theta) - 120(\cos\theta)^3 + 432(\cos\theta)^5 - 576(\cos\theta)^7 + 256(\cos\theta)^9$
$\cos 10\theta$	$-1 + 50(\cos\theta)^2 - 400(\cos\theta)^4 + 1120(\cos\theta)^6 - 1280(\cos\theta)^8 + 512(\cos\theta)^{10}$
$\cos 11\theta$	$-11(\cos\theta) + 220(\cos\theta)^3 - 1232(\cos\theta)^5 + 2816(\cos\theta)^7 - 2816(\cos\theta)^9 + 1024(\cos\theta)^{11}$
$\cos 12\theta$	$1 - 72(\cos\theta)^2 + 840(\cos\theta)^4 - 3584(\cos\theta)^6 + 6912(\cos\theta)^8 - 6144(\cos\theta)^{10} + 2048(\cos\theta)^{12}$

Chebyshev Polynomials

As is the custom in mathematics, once we have an interesting new object, we look at it as something to study in its own right.

Many systems have them built in.

```
F1 Tools | F2 Control | F3 I/O | F4 Var | F5 Find... | F6 Mode
:t(n)
:Func
:If n=0 Then
:Return 1
:ElseIf n=1 Then
:Return x
:Else
:Return 2*x*t(n-1)-t(n-2)
:EndIf
MAIN    RAD AUTO    FUNC
```

Any CAS that allows you to model recursively defined functions can generate Chebyshev polynomials. In *Mathematica*, three lines will do it:

```
t[0] := 1
t[1] := x
t[n_] := 2x t[n-1] - t[n-2]
```

The TI-89 requires that you package all the cases statements into a single function:

```
t(n)
Func
If n=0 Then
Return 1
ElseIf n=1 Then
Return x
Else
Return 2*x*t(n-1)-t(n-2)
EndIf
Endfunc
```

Once you can generate the polynomials, you can look for patterns and try to see why they hold. Here are some Chebyshev polynomials:

Chebyshev polynomials	
1	x
2	$-1+2x^2$
3	$-3x+4x^3$
4	$1-8x^2+8x^4$
5	$5x-20x^3+16x^5$
6	$-1+18x^2-48x^4+32x^6$
7	$-7x+56x^3-112x^5+64x^7$
8	$1-32x^2+160x^4-256x^6+128x^8$
9	$9x-120x^3+432x^5-576x^7+256x^9$
10	$-1+50x^2-400x^4+1120x^6-1280x^8+512x^{10}$
11	$-11x+220x^3-1232x^5+2816x^7-2816x^9+1024x^{11}$
12	$1-72x^2+840x^4-3584x^6+6912x^8-6144x^{10}+2048x^{12}$
13	$13x-364x^3+2912x^5-9984x^7+16640x^9-13312x^{11}+4096x^{13}$
14	$-1+98x^2-1568x^4+9408x^6-26880x^8+39424x^{10}-28672x^{12}+8192x^{14}$
15	$-15x+560x^3-6048x^5+28800x^7-70400x^9+92160x^{11}-61440x^{13}+16384x^{15}$
16	$1-128x^2+2688x^4-21504x^6+84480x^8-180224x^{10}+212992x^{12}-131072x^{14}+32768x^{16}$

And here are their factored forms:

	Factored Chebyshev polynomials
1	x
2	$-1+2x^2$
3	$x(-3+4x^2)$
4	$1-8x^2+8x^4$
5	$x(5-20x^2+16x^4)$
6	$(-1+2x^2)(1-16x^2+16x^4)$
7	$x(-7+56x^2-112x^4+64x^6)$
8	$1-32x^2+160x^4-256x^6+128x^8$
9	$x(3-4x^2)(3-36x^2+96x^4-64x^6)$
10	$(-1+2x^2)(1-48x^2+304x^4-512x^6+256x^8)$
11	$x(-11+220x^2-1232x^4+2816x^6-2816x^8+1024x^{10})$
12	$(1-8x^2+8x^4)(1-64x^2+320x^4-512x^6+256x^8)$
13	$x(13-364x^2+2912x^4-9984x^6+16640x^8-13312x^{10}+4096x^{12})$
14	$(-1+2x^2)(1-96x^2+1376x^4-6656x^6+13568x^8-12288x^{10}+4096x^{12})$
15	$x(3-4x^2)(-5+20x^2-16x^4)(1-32x^2+224x^4-448x^6+256x^8)$
16	$1-128x^2+2688x^4-21504x^6+84480x^8-180224x^{10}+212992x^{12}-131072x^{14}+32768x^{16}$

And some graphs:

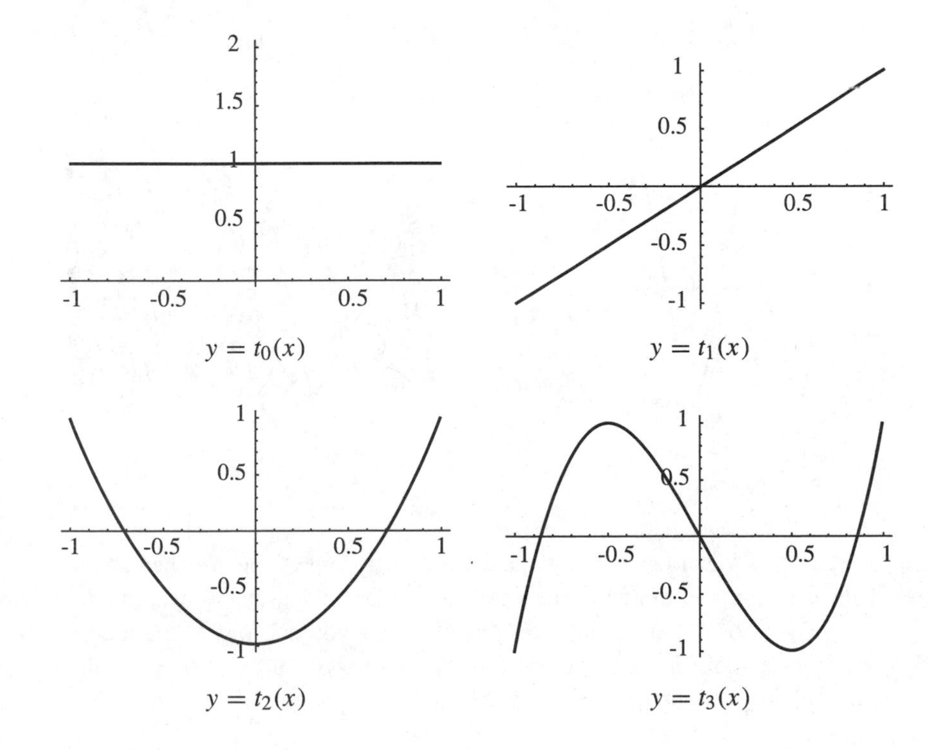

$y = t_0(x)$ $y = t_1(x)$

$y = t_2(x)$ $y = t_3(x)$

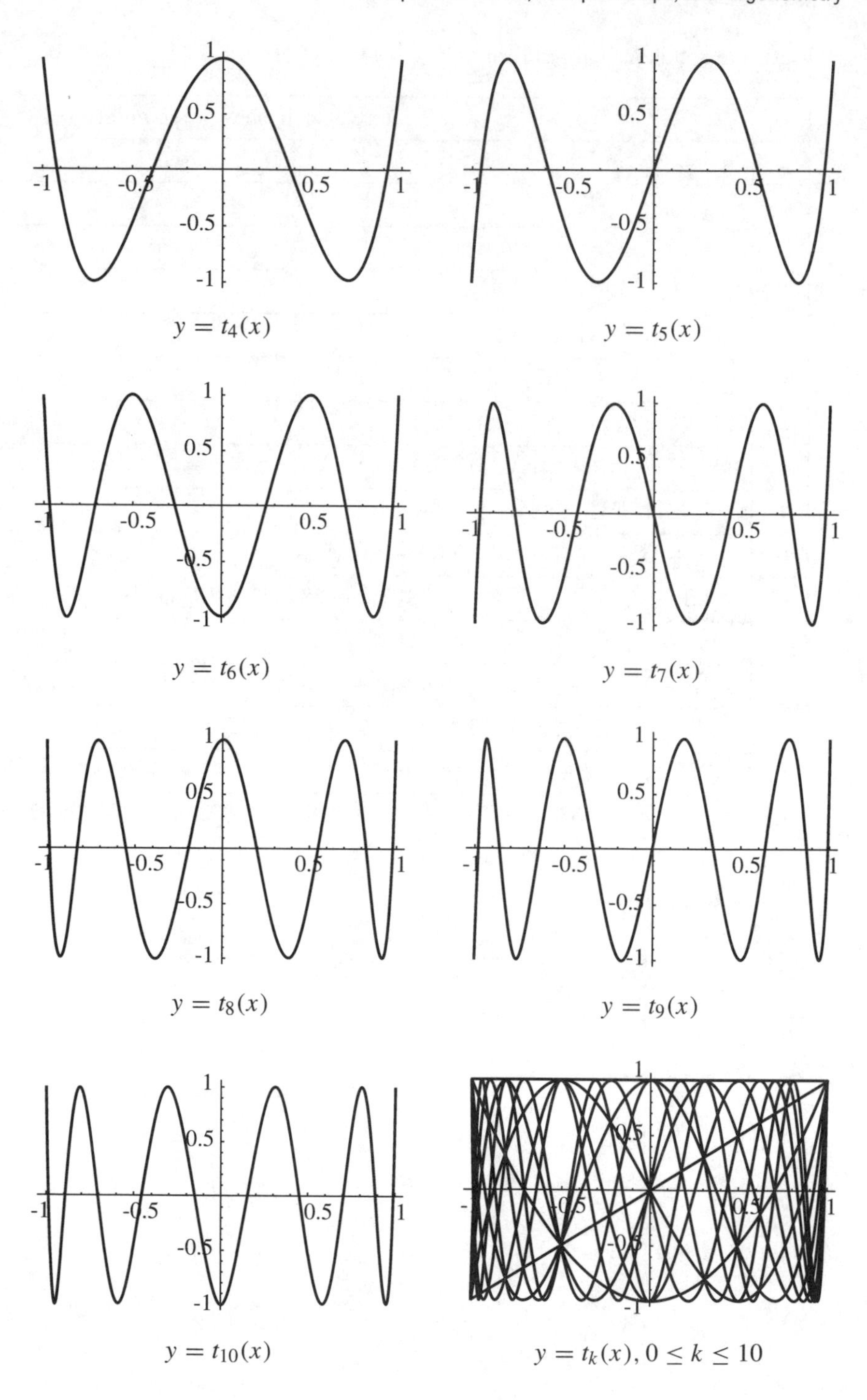

$y = t_4(x)$ $y = t_5(x)$

$y = t_6(x)$ $y = t_7(x)$

$y = t_8(x)$ $y = t_9(x)$

$y = t_{10}(x)$ $y = t_k(x), 0 \leq k \leq 10$

Well, as a friend used to say, there's a candy store of patterns here. Some of them will be developed in the problems, and you'll probably want to find some of your own. Let's look at one, just to show you one way to think about the proofs: It looks from the graphs like the set of zeros of the t_k (the so-called *spectrum* of t_k) lies between -1 and 1. Let's see why.

Start with an example, say $t_3(x) = -3x + 4x^3$. Of course, we can just find the zeros by factoring, but let's look for a method that's general in principle. It's a cubic, and it seems (from its graph) to have three real zeros. Can we produce them? Well, by Theorem 7,

If $-3x + 4x^3 = 0$, $x(-3 + 4x^2) = 0$, so $x = 0$ or $x^2 = \frac{3}{4}$.

$$t_3(\cos\theta) = \cos 3\theta.$$

So, we can find a root if we can find a value of θ such that $\cos 3\theta = 0$; the root will then be $\cos\theta$. But there are plenty of angles θ that make $\cos 3\theta = 0$. They are

Cosine is 0 precisely at the odd multiples of $\frac{\pi}{2}$. That's problem 58.

$$3\theta = \cdots - \frac{3\pi}{2}, -\frac{\pi}{2}, \frac{\pi}{2}, \frac{3\pi}{2}, \frac{5\pi}{2} \cdots$$

or (dividing all these by 3),

$$\theta = \cdots - \frac{\pi}{2}, -\frac{\pi}{6}, \frac{\pi}{6}, \frac{\pi}{2}, \frac{5\pi}{6} \cdots$$

But then (taking cosines of these angles)

$$\cos\theta = \cdots 0, \frac{\sqrt{3}}{2}, \frac{\sqrt{3}}{2}, 0, -\frac{\sqrt{3}}{2} \cdots$$

There: We have three zeros of t_3. They are

$$0, \frac{\sqrt{3}}{2}, -\frac{\sqrt{3}}{2}.$$

Oh, and these are all between -1 and 1 because they are *cosines* of angles, and cosines are always between -1 and 1.

The idea is the same in the general case—we produce k distinct zeros for t_k and these k zeros are each cosines of angles, so they are between -1 and 1. Theorem 7 tells us that $t_k(\cos\theta) = \cos k\theta$, so, to make t_k produce 0, feed it an input of $\cos\theta$ where $\cos k\theta = 0$. We'll find k distinct such inputs using the following table:

t_k has degree k by problem 66. So, it has k zeros.

	$\longleftarrow$ There are k columns to the right of the vertical bar $\longrightarrow$					
$k\theta$ so that $\cos(k\theta) = 0$	$\frac{\pi}{2}$	$3 \cdot \frac{\pi}{2}$	$5 \cdot \frac{\pi}{2}$	$7 \cdot \frac{\pi}{2}$	$\cdots$	$(2k-1)\frac{\pi}{2}$
θ	$\frac{\frac{\pi}{2}}{k}$	$\frac{3\cdot\frac{\pi}{2}}{k}$	$\frac{5\cdot\frac{\pi}{2}}{k}$	$\frac{7\cdot\frac{\pi}{2}}{k}$	$\cdots$	$\frac{(2k-1)\frac{\pi}{2}}{k}$
$\cos\theta$	$\cos\left(\frac{\frac{\pi}{2}}{k}\right)$	$\cos\left(\frac{3\cdot\frac{\pi}{2}}{k}\right)$	$\cos\left(\frac{5\cdot\frac{\pi}{2}}{k}\right)$	$\cos\left(\frac{7\cdot\frac{\pi}{2}}{k}\right)$	$\cdots$	$\cos\left(\frac{(2k-1)\frac{\pi}{2}}{k}\right)$

The numbers in the $\cos\theta$ row all make t_k produce 0, because

$$t_k\left(\cos\left(\frac{\text{odd} \times \frac{\pi}{2}}{k}\right)\right) = \cos k\left(\frac{\text{odd} \times \frac{\pi}{2}}{k}\right) = \cos\left(\text{odd} \times \frac{\pi}{2}\right) = 0.$$

So, we have k numbers α such that $t_k(\alpha) = 0$. And they are all between -1 and 1 (because they are cosines). And t_k has degree k, so it has at most k distinct zeros. So, if these numbers are all distinct, we know that all the roots of t_k are between -1 and 1.

Problem 73: If $\cos A = \cos B$, then either $A + B$ or $A - B$ is an integer multiple of 2π.

Well, they *are* distinct, thanks to problem 73. The argument goes like this: Each of the numbers we produced as a zero of t_k is of the form

$$\cos\left(\frac{(2j-1)\cdot\frac{\pi}{2}}{k}\right) \qquad \text{where } j \text{ is some integer between 1 and } 2k-1.$$

If two of these are equal, say

$$\cos\left(\frac{(2p-1)\cdot\frac{\pi}{2}}{k}\right)$$

and

$$\cos\left(\frac{(2q-1)\cdot\frac{\pi}{2}}{k}\right),$$

with $p \neq q$, problem 73 tells us that either

$$\frac{(2p-1)\cdot\frac{\pi}{2}}{k} = 2\pi r + \frac{(2q-1)\cdot\frac{\pi}{2}}{k} \qquad \text{for some integer } r, \text{ or}$$

$$\frac{(2p-1)\cdot\frac{\pi}{2}}{k} = 2\pi r - \frac{(2q-1)\cdot\frac{\pi}{2}}{k} \qquad \text{for some integer } r.$$

Multiply both sides by k and we'd have either

$$(2p-1)\cdot\frac{\pi}{2} = 2\pi kr + (2q-1)\cdot\frac{\pi}{2} \qquad \text{for some integer } r, \text{ or}$$

$$(2p-1)\cdot\frac{\pi}{2} = 2\pi kr - (2q-1)\cdot\frac{\pi}{2} \qquad \text{for some integer } r.$$

One more simplification gives:

$$p - q = 2kr \qquad \text{for some integer } r, \text{ or}$$

$$p + q = 2kr \qquad \text{for some integer } r.$$

If p and q are distinct and between 1 and $2k - 1$, neither their difference nor their sum can be an integer multiple of $2k$. So, we have produced k distinct roots for t_k, all between -1 and 1. After all this work, let's summarize all this in a theorem:

Theorem 8. *t_k has k distinct zeros, all between -1 and 1. In fact, they are*

$$\left\{\cos\left(\frac{(2j-1)\cdot\frac{\pi}{2}}{k}\right) \mid 1 \leq j \leq 2k-1\right\}.$$

This is a perfect example of where form and function come together for polynomials.

So, one basic strategy when looking to prove results about Chebyshev polynomials is to exploit the fact that they give multiple angle formulas for cosine.

Problems

On page 60, a picture illustrates some trigonometric identities based on the definitions of sine and cosine in the second quadrant. Here is a picture for quadrant 3.

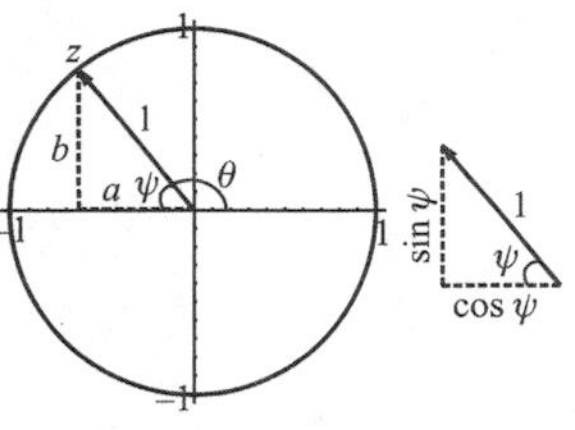

$\cos\theta = -\cos(\pi - \theta)$
$\sin\theta = \sin(\pi - \theta)$

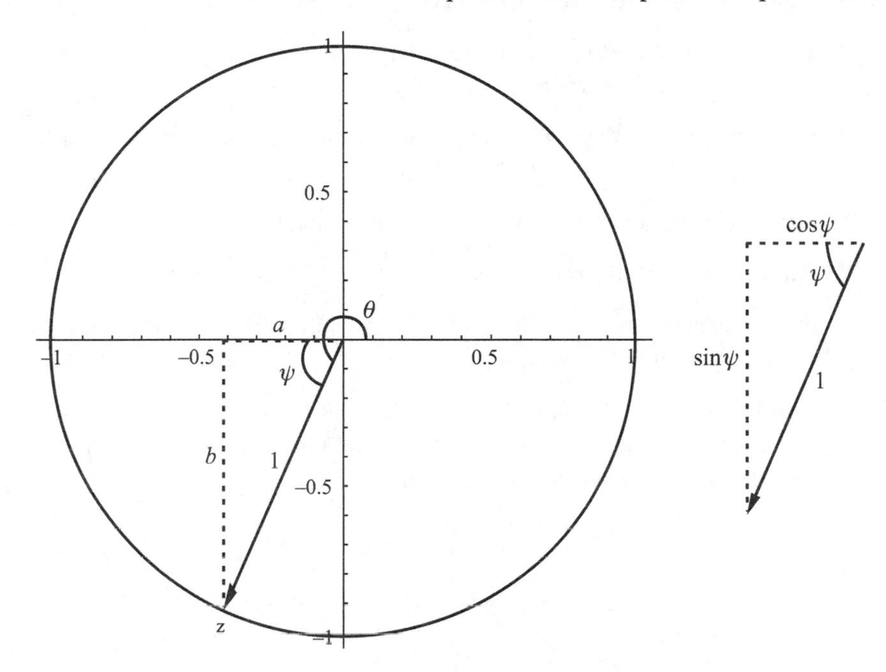

55. **(a)** What is the relationship between ψ and θ?
(b) What is the relationship between $\cos\psi$ and $\cos\theta$?
(c) What is the relationship between $\sin\psi$ and $\sin\theta$?
(d) Come up with two more trigonometric identities—one each for sine and cosine.

56. Draw the picture for quadrant 4 and prove two more identities—one each for sine and cosine.

57. Use the "back to the definition" approach to make a table of *exact* values for sine and cosine that contains at least four angles in each quadrant.

Think about angles like $\frac{\pi}{6}$, $\frac{\pi}{2}$, and $\frac{\pi}{4}$.

58. Show that if $\cos\theta = 0$, θ is (in radians) an odd integer multiple of $\frac{\pi}{2}$.

59. Show that if $\sin\theta = 0$, θ is (in radians) an integer multiple of π.

An integer multiple of 180° is an *even* integer multiple of 90°.

60. Find a function that outputs zero at every integer and nowhere else.

61. Find a function that agrees with $f(x) = 2x+1$ at every integer and nowhere else.

This would *really* fool the test makers (see Chapter 1).

62. **Prove, or Disprove and Salvage:** For each part, decide if the identity is true or false. If it is true, draw a picture to illustrate it and prove it. If it is false, explain why (find a counterexample!). If you decide it is false, can you create a slightly different identity that is true and prove that one?

Try problems 62 and 63 using both methods: "go back to the definitions" and "use multiplication in $\mathbb{C}$."

(a) $\cos(-\theta) = -\cos(\theta)$ **(b)** $\cos(-\theta) = -\cos(\theta)$
(c) $\cos(\frac{\pi}{2} - \theta) = \sin\theta$ **(d)** $\cos(\frac{\pi}{2} + \theta) = \sin\theta$

63. Express $\sin(\frac{\pi}{2} + \theta)$ in terms of $\sin\theta$ and $\cos\theta$.

64. Express $\sin 3\theta$ in terms of $\sin\theta$.

65. Express $\sin 4\theta$ in terms of $\sin\theta$ and $\cos\theta$.

66. Show that the degree of the Chebyshev polynomial t_k is k. Find a formula for the coefficient for x^k in t_k.

67. Find a formula for the coefficient of some term in t_k other than the highest degree one, x^k.

68. State and prove a fact about Chebyshev polynomials.

69. Express $\cos(\theta/2)$ in terms of $\cos\theta$.

70. Find an expression for $\cos 18°$ and $\cos 54°$ that involves only the four operations of arithmetic and square roots.

71. Find an expression for $\cos 72° + i \sin 72°$ that involves only the four operations of arithmetic and square roots. Use this to express the roots of $x^5 - 1 = 0$ without any sines and cosines.

72. Show that

$$\cos A - \cos B = -2 \sin \tfrac{1}{2}(A + B) \sin \tfrac{1}{2}(A - B).$$

If $\cos A = \cos B$, $\cos A - \cos B = 0$.

73. Suppose $\cos A = \cos B$. Show that either

$$A = 2j\pi + B \quad \text{for some integer } j\text{, or}$$
$$A = 2j\pi - B \quad \text{for some integer } j.$$

These identities have gotten a bad name among teachers lately. Why? They seem perfectly nice.

74. Use the method of problem 72 to show that:

(a) $\cos A + \cos B = 2 \cos \frac{1}{2}(A + B) \cos \frac{1}{2}(A - B)$,
(b) $\sin A + \sin B = 2 \sin \frac{1}{2}(A + B) \cos \frac{1}{2}(A - B)$,
(c) $\sin A - \sin B = -2 \cos \frac{1}{2}(A + B) \sin \frac{1}{2}(A - B)$.

Your students will love you for this.

Problems 75–77 develop a method for inventing trigonometric identities for your students.

75. Just for fun, prove the following old-fashioned trigonometric identities.

(a) $$\cos^2 x + 2\cos^3 x \sin x + 2\cos x \sin^3 x + \sin^4 x = \cos^4 x + 2\cos x \sin x + \sin^2 x$$

(b) $$\cos^2 x + \cos x \sin x + \sin^4 x = \cos^4 x + \cos^3 x \sin x + \sin^2 x - \cos x \sin^3 x$$

76. The identity in problem 75(a) was invented by the following process:

(a) Expand $(C^2 + S^2 - 1)(S^2 + 2CS - C^2)$.

(b) Set this expression equal to 0 and move all the negative terms to the other side of the equation.

(c) Replace C by $\cos x$ and S by $\sin x$.

Why does this produce a valid trigonometric identity?

77. Explain why the following algorithm produces valid trigonometric identities.

(a) Take a polynomial $f(S, C)$ in the two variables S and C.

(b) Thinking of C as the primary variable, divide $f(S, C)$ by $C^2 + S^2 - 1$. This will produce a remainder of the form

$$r(S, C) = h(S) \cdot C + j(S),$$

where $h(S)$ and $j(S)$ are polynomials in S.

(c) Then $f(\sin x, \cos x) = r(\sin x, \cos x)$ is an identity.

You get another identity if you think of S as the primary variable.

In *Mathematica*, you can use `PolynomialRemainder[f(S, C), C^2 + S^2 - 1, S]` to calculate $r(S, C)$.

78. State and prove a theorem that describes the symmetry properties of the graphs of the Chebyshev polynomials.

79. If k and m are positive integers, show that

$$t_k(\cos(m\theta)) = \cos(km\theta).$$

80. Show that the Chebyshev polynomials commute in the sense that

$$t_k\big(t_\ell(x)\big) = t_\ell\big(t_k(x)\big)$$

for all nonnegative integers k and l.

Can you think of other polynomials that commute in this way?

81. Develop a rule that lets you tell when the roots of $t_m(x) = 0$ are also roots of $t_n(x) = 0$. Use this to state and prove a theorem that tells when $t_m(x)$ is a factor of $t_n(x)$.

Here, "factor" means there is a polynomial $f(x)$ with integer coefficients such that $t_n(x) = f(x)t_m(x)$.

82. Theorem 8 shows that all the zeros of t_k are between -1 and 1. Show that as $k \to \infty$, the largest zero approaches 1 and the smallest zero approaches -1.

83. It's rumored that there's an explicit formula for t_k:

$$t_k(x) \stackrel{??}{=} \sum_{j=0}^{\infty} (-1)^j \binom{k}{2j} x^{k-2j}(1 - x^2)^j.$$

This is really a finite sum, because the $\binom{k}{2j}$ are 0 when $j > \lfloor k/2 \rfloor$, the greatest integer less than or equal to $k/2$.

Write out (or use a CAS to calculate) the first few t_k (according to this formula) to see if they agree with the table on page 129. Is the rumor true? If so, prove it. If not, find an example where it fails.

84. Develop a sequence of polynomials u_k so that

$$\sin k\theta = \sin\theta \cdot u_k(\cos\theta).$$

When is $\sin k\theta$ simply a polynomial in $\sin\theta$?

85. It's rumored that there's another explicit formula for t_k:

$$t_k(x) = \frac{(x + \sqrt{x^2 - 1})^n + (x - \sqrt{x^2 - 1})^n}{2}.$$

How could such a rumor have gotten started? Is it true?

3.5 Complex Maps

The geometric model for multiplication and addition of complex numbers allows us to think about what the graphs of complex functions might look like. Of course, you can't use the x, y-plane to model a complex function, since the whole plane is just taken up with representing the numbers themselves. You need to be a little more creative.

Method 1: Look at Regions

One way to picture complex functions is to look at what happens to a particular region or set of points when you apply the function. In this case, you look at two different planes: one plane with a region that shows the domain of the function, and another that shows the image of that region when the function is applied.

Example: What will happen to these regions when you apply the function

$$f(z) = z + (1 + i)?$$

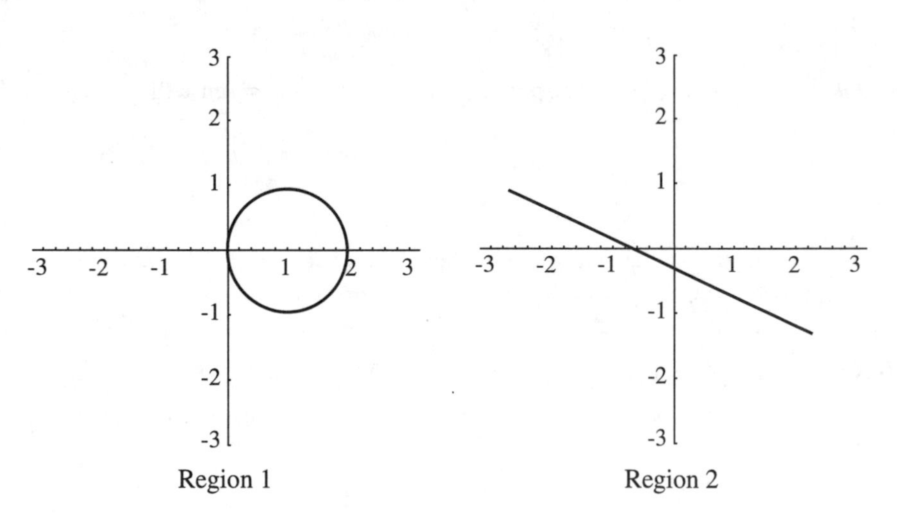

Well, you can try several key points, and get a sense of what will happen for the whole region.

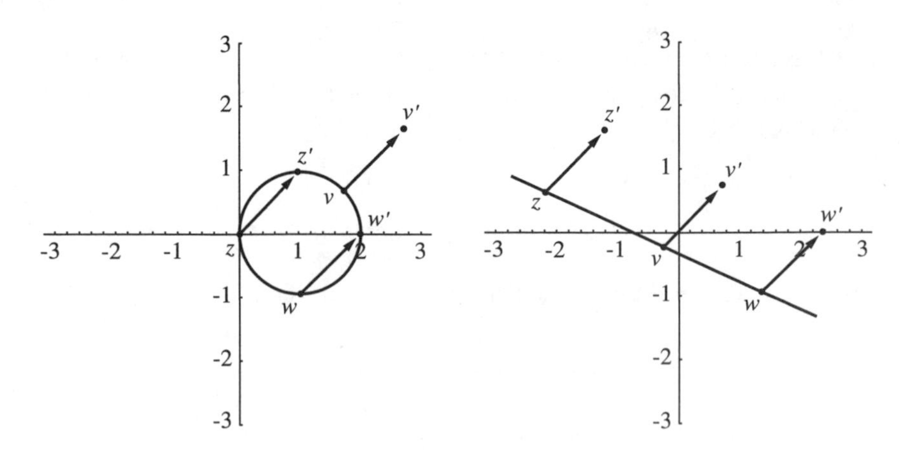

It looks as if each point is translated up one unit and over one unit. So we can plot the images of each region.

This makes sense, if you think back to our geometric interpretation of adding complex numbers.

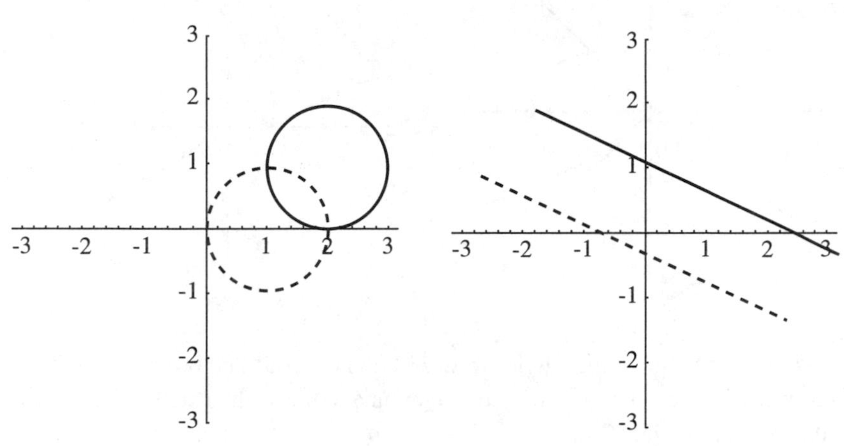

Image of Region 1 Image of Region 2

What will happen to these regions when you apply the function

$$f(z) = 2\left(\cos\frac{\pi}{4} + i\sin\frac{\pi}{4}\right)\cdot z?$$

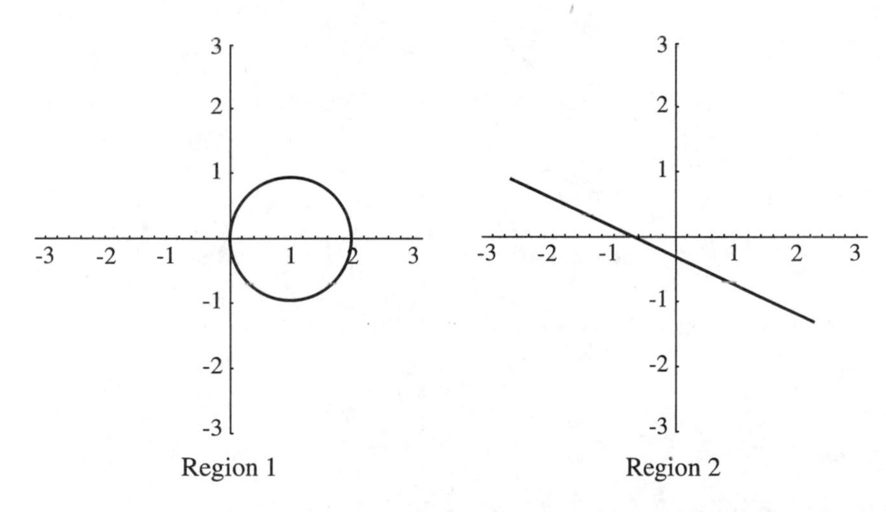

Region 1 Region 2

Again, try several key points, and get a sense of what will happen for the whole region. For each point, it will double in length and rotate by 45°.

Remember, 45° is the same rotation as $\frac{\pi}{4}$ radians.

Are z', w', and v' on a line?

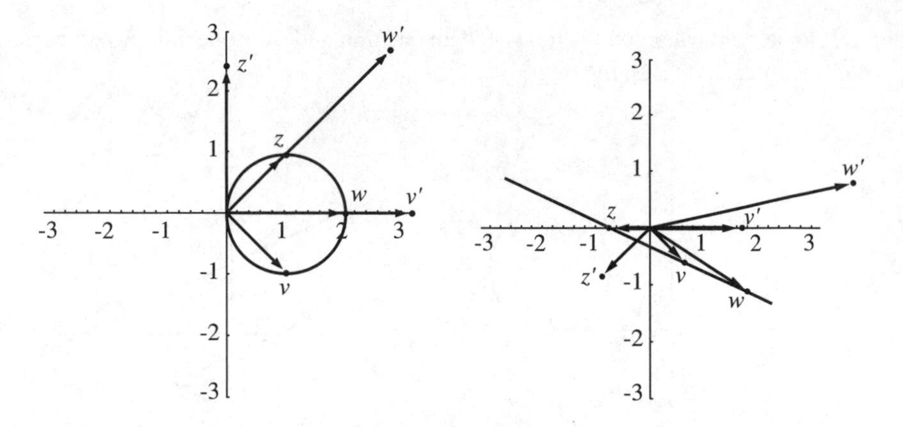

For the first region, the whole circle will rotate by 45° and be scaled by 2, but remember that the center of both the rotation and the scaling will be the origin. Likewise for the line.

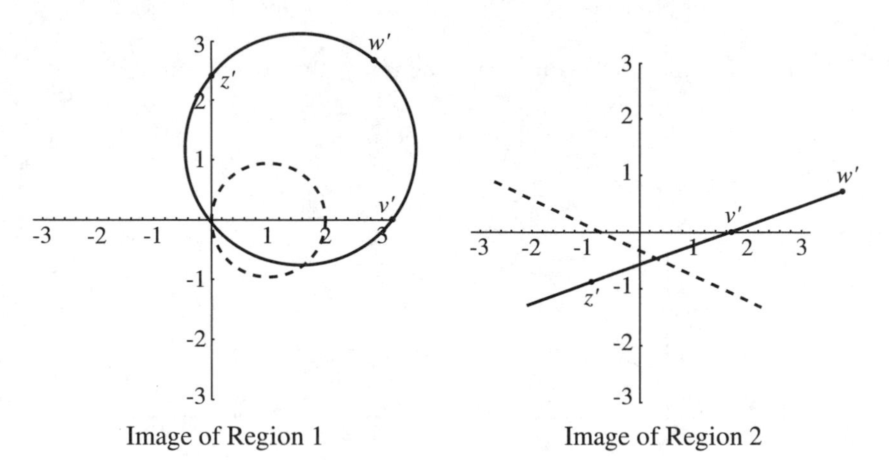

Image of Region 1 Image of Region 2

What will happen to this when you apply the function

$$f(z) = z^2?$$

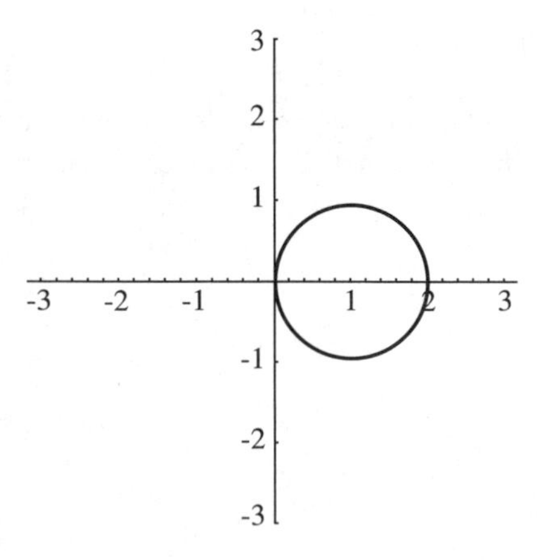

Region 1

Again, try several key points, and get a sense of what will happen for the whole region. What happens to each point depends on the point itself. The distance from the origin is squared, and the angle is doubled. Here's how you can think about it. If $z = r\,(\cos\theta + i\sin\theta)$, then

$$f(z) = z^2 = r^2\big(\cos(\theta+\theta)i\sin(\theta+\theta)\big) = r^2(\cos 2\theta + i\sin 2\theta).$$

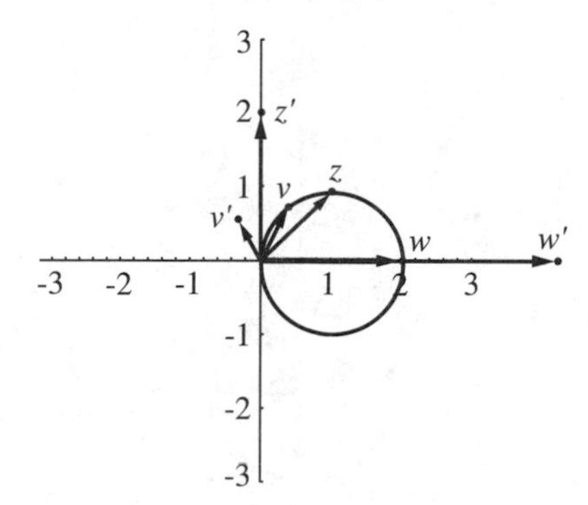

FIGURE 1

In this case, it's not as easy to tell what will happen to the whole region, since the points don't all behave the same way. You can think about pieces, though.

For example, for the top half of the circle, the angles range from 0° to 90°. The image of those points will have angles that range from 0° to 180°, so the points between the origin and the point marked z in Figure 1 must map to the second quadrant.

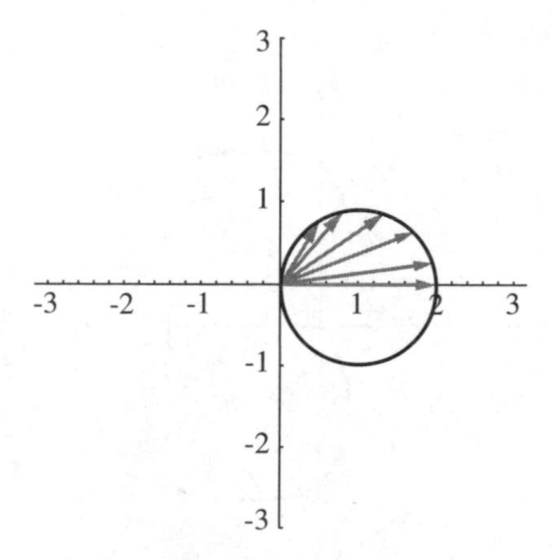

Angles range from 0° to 90°

Similarly, the points in the lower half of the circle have angles from 270° to 360°, so the image points will be from 540° (which is equivalent to 180°) to 720° (which is equivalent to 360°).

Also, some points will move closer to the origin, and some will move farther away. In particular, if $|z| < 1$, then z^2 will be closer to the origin than z. When does this happen for points on the circle? We can do some algebra to figure it out. For now, it will help to think of z in the form $x + iy$. The equation for the circle is $(x-1)^2 + y^2 = 1$, and $|z| = x^2 + y^2$. Now, the circles with equations $x^2 + y^2 = 1$ and $(x-1)^2 + y^2 = 1$ intersect at $(\frac{1}{2}, \pm\frac{\sqrt{3}}{2})$, as in the picture below.

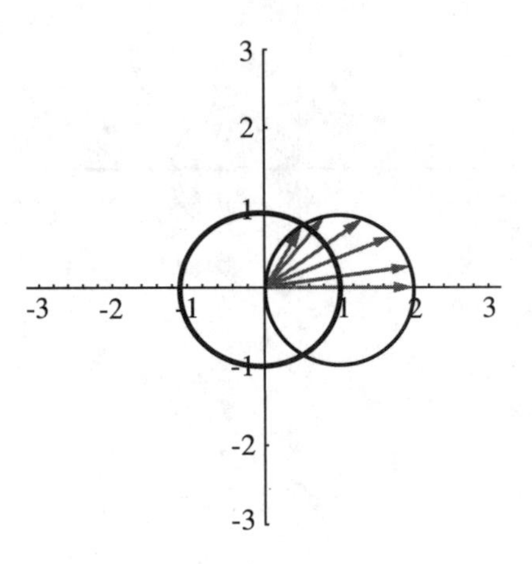

Notice how we move back and forth between the complex plane and the Cartesian plane. That happens a lot when thinking about complex functions.

So for values of x between 0 and $\frac{1}{2}$, the image points are closer to the origin. For $x = \frac{1}{2}$, $|f(z)| = |z| = 1$. And for $x > \frac{1}{2}$, the image points move farther from the origin.

Putting these pieces of information together with the points we originally plotted, we can get a good sense for what the image looks like.

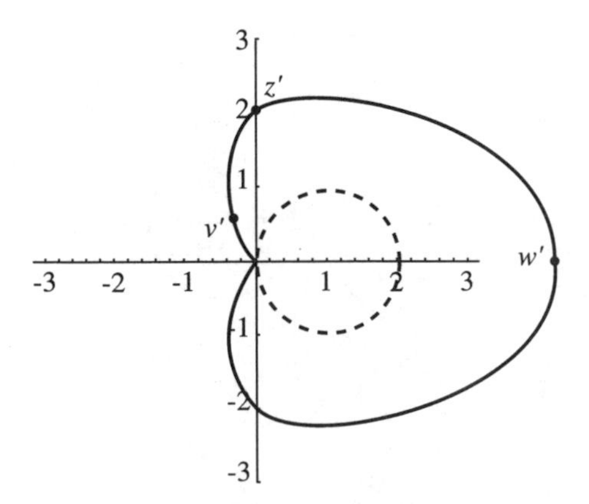

Image of Region 1

So, one way to picture functions on complex numbers—to draw their graphs—is to draw a "domain" plane with some region in it and a "range" plane with the image of that region when the function is applied.

Method 2: Look at Iterates

Another way to think about what happens with complex functions is to look at what happens to one particular point as the function is iterated. Investigating the iteration of complex functions is an important tool in the study of *dynamical systems*.

In a dynamical system, what happens at the next stage depends on the current stage. For example, in studying population dynamics, next year's population depends on this year's population—how many animals reproduce, compete for resources, etc.

Take a function, say $g(z) = \frac{1}{2}(\cos\frac{\pi}{4} + i\sin\frac{\pi}{4}) \cdot z$, and start with an initial value of z, called the **seed**. You can generate a sequence of iterates of the function. Suppose $z = 1$ is our seed:

Note that $g^2(z)$ means $g(g(z))$, or the second iterate. The exponent tells you how many times to apply the function.

$$\begin{aligned}
g(1) &= \frac{1}{2}\left(\cos\frac{\pi}{4} + i\sin\frac{\pi}{4}\right) \cdot 1 \\
&= \frac{1}{2}\left(\cos\frac{\pi}{4} + i\sin\frac{\pi}{4}\right), \\
g^2(1) &= g(g(1)) \\
&= \left(\frac{1}{2}\left(\cos\frac{\pi}{4} + i\sin\frac{\pi}{4}\right)\right) \cdot \left(\frac{1}{2}\left(\cos\frac{\pi}{4} + i\sin\frac{\pi}{4}\right)\right) \\
&= \frac{1}{4}\left(\cos\frac{\pi}{2} + i\sin\frac{\pi}{2}\right) \\
&= \frac{1}{4}i, \\
g^3(1) &= g\left(g^2(1)\right) \\
&= \left(\frac{1}{2}\left(\cos\frac{\pi}{4} + i\sin\frac{\pi}{4}\right)\right) \cdot \frac{1}{4}i \\
&= \left(\frac{1}{2}\left(\cos\frac{\pi}{4} + i\sin\frac{\pi}{4}\right)\right)\left(\frac{1}{4}\left(\cos\frac{\pi}{2} + i\sin\frac{\pi}{2}\right)\right) \\
&= \frac{1}{8}\left(\cos\frac{3\pi}{4} + i\sin\frac{3\pi}{4}\right)
\end{aligned}$$

and, in general

$$g^n(1) = \left(\frac{1}{2}\right)^n \left(\cos\frac{n\pi}{4} + i\sin\frac{n\pi}{4}\right).$$

When you start with a seed of z_0, the sequence of iterates is called the **orbit** of z_0. You can graph this sequence of iterates: $z_0, z_1 = g(z_0), z_2 = g(z_1), \ldots$ to see what the orbit looks like. Does it bounce back and forth? Spiral around? Move closer to or farther from the origin?

Example: Sketch the orbit for 1 of the function $g(z) = \frac{1}{2}\left(\cos\frac{\pi}{4} + i\sin\frac{\pi}{4}\right) \cdot z$. We've already done the first few computations above, but to graph the orbit you could really just think to yourself, "Each application of the function shrinks the vector by $\frac{1}{2}$ and rotates it by $\frac{\pi}{4}$ (45°)." The orbit seems to spiral in towards the origin.

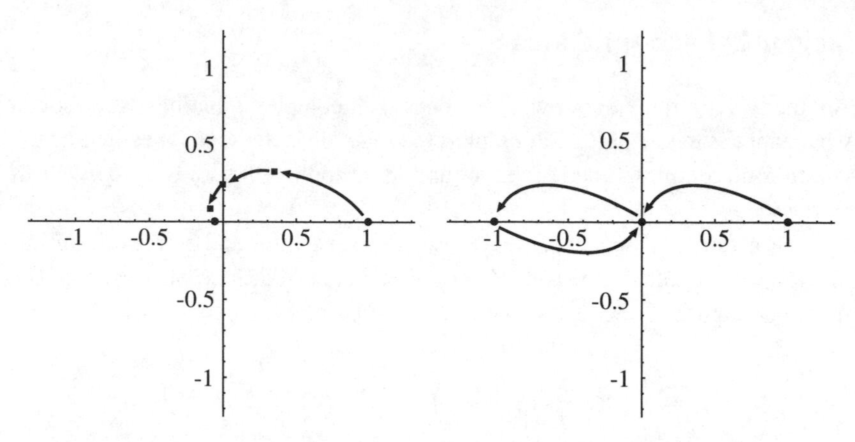

$$h(1) = 1^2 - 1 = 0,$$
$$h(0) = 0^2 - 1 = -1,$$
$$h(-1) = (-1)^2 - 1 = 0,$$
$$h(0) = 0^2 - 1 = -1,$$
$$\vdots$$

Suppose you want to sketch the orbit for 1 of the function $h(z) = z^2 - 1$. In this case, it's not obvious what the function does at each step, so you may want to do the computations to help you picture it. In this case, the orbit flips back and forth between $z = -1$ and $z = 0$.

Problems

Problems 86–88 ask you to investigate complex maps by looking at the images of specific regions in the plane.

86. Sketch what happens to each region below when you apply the function

(a) $f(z) = z + (-1 - 2i)$ **(b)** $f(z) = \overline{z}$

(c) $f(z) = 1/z$

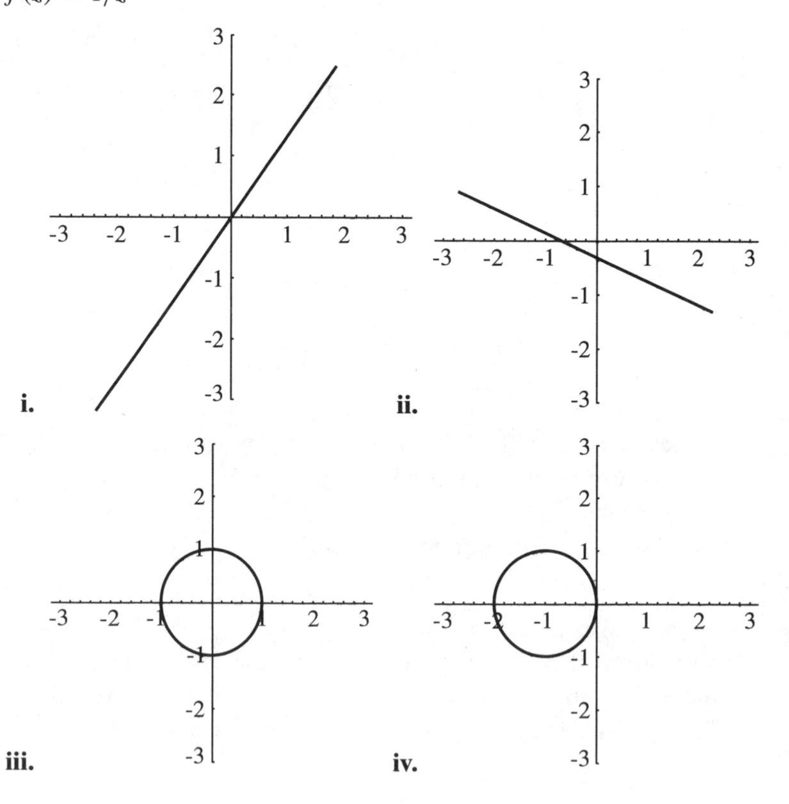

87. Sketch what happens to each region below when you apply the function

(a) $f(z) = \frac{1}{2}\left(\cos\frac{\pi}{2} + i\sin\frac{\pi}{2}\right)\cdot z$ **(b)** $f(z) = \left(\cos\frac{\pi}{2} + i\sin\frac{\pi}{2}\right)\cdot z$

(c) $f(z) = 2\left(\cos\frac{\pi}{2} + i\sin\frac{\pi}{2}\right)\cdot z$

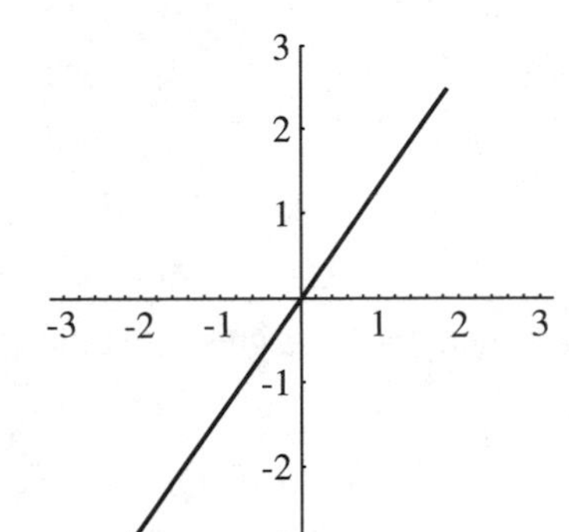

i.

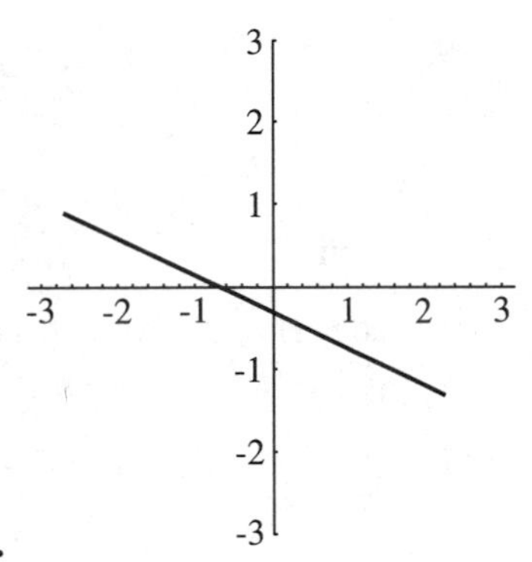

ii.

iii.

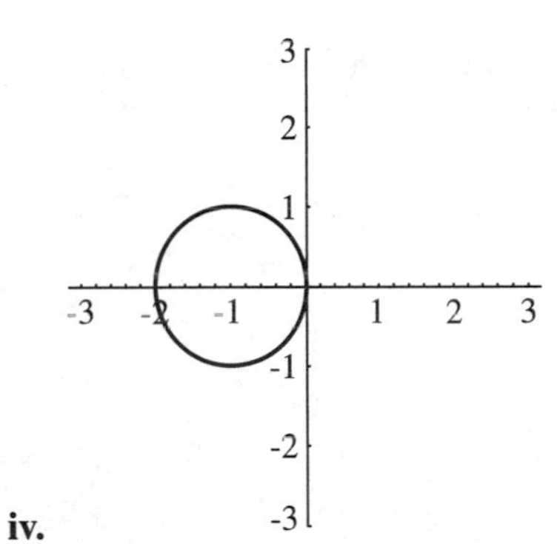

iv.

88. Sketch what happens to each region below when you apply the function

(a) $f(z) = z^2$ **(b)** $f(z) = \sqrt{z}$

(c) $f(z) = z^2 + (1+i)$ **(d)** $f(z) = z^2 + z$

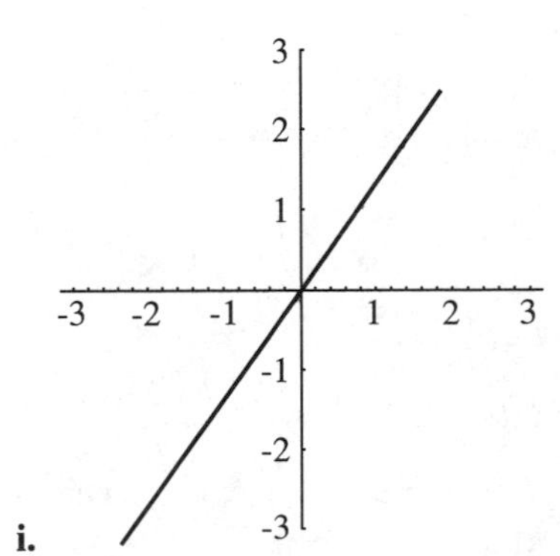

i.

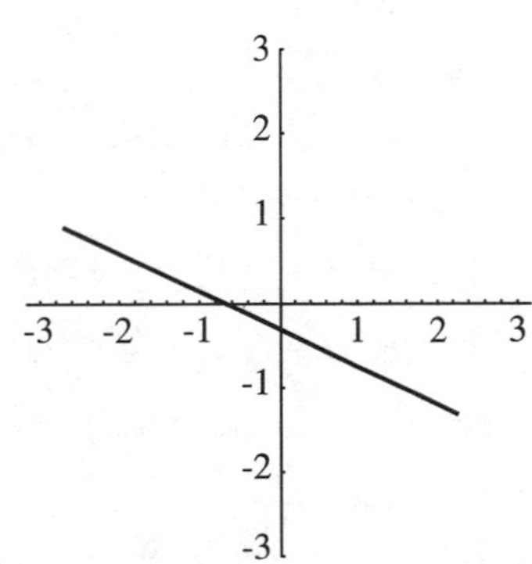

ii.

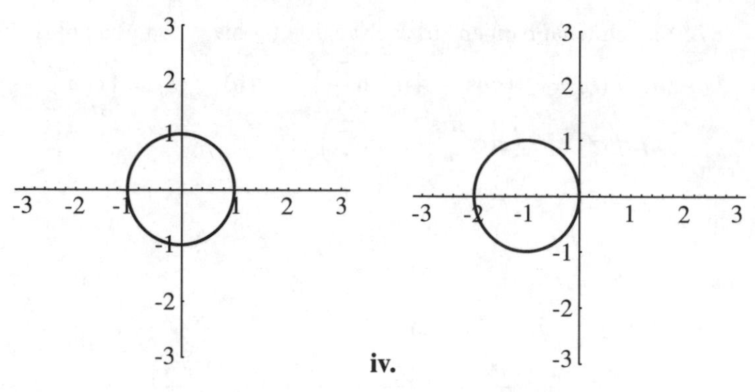

iii. iv.

89. Let $f(z) = z^2$. Find a set $S \subset \mathbb{C}$ with the following properties:

(a) $i \in S$ **(b)** if $z \in S$, $f(z) \in S$

90. Let $f(z) = z^2$. Find a set $S \subset \mathbb{C}$ with the following properties:

(a) $\cos 20° + i \sin 20° \in S$ **(b)** if $z \in S$, $f(z) \in S$

For problems 91–93, sketch and describe the general behavior of the orbit for these three points. Find or estimate at least four points on the orbit.

(a) $z = 1 + i = \sqrt{2}(\cos 45° + i \sin 45°)$
(b) $z = -i = \cos 270° + i \sin 270°$
(c) $z = -\frac{1}{2} - \frac{\sqrt{3}}{4} i = \frac{1}{2}\,(\cos 300° + i \sin 300°)$

All angle measures are in degrees.

91. $f(z) = \frac{1}{3}\,(\cos 30° + i \sin 30°) \cdot z$

92. $g(z) = (\cos 72° + i \sin 72°) \cdot z$

93. $h(z) = 3(\cos 270° + i \sin 270°) \cdot z$

94. $b(z) = \overline{z}$

95. Find the fixed points of $f(z) = z^2$. That is, find all complex numbers w so that $f(w) = w$.

96. Consider the function $f(z) = \alpha \cdot z$. Sketch the orbit of $z = 1$ for each of the following values of α. What do all of the orbits have in common? How are they different?

(a) $\alpha = \cos \frac{\pi}{4} + i \sin \frac{\pi}{4}$ **(b)** $\alpha = \cos\left(\frac{2\pi}{13}\right) + i \sin\left(\frac{2\pi}{13}\right)$
(c) $\alpha = \cos\left(\frac{\pi\sqrt{5}}{4}\right) + i \sin\left(\frac{\pi\sqrt{5}}{4}\right)$
(d) $\alpha = \cos(100\sqrt{2})° + i \sin(100\sqrt{2})°$

97. Consider functions of the form $f(z) = \alpha \cdot z$ where α is a complex number. Sketch and/or describe the orbit of $z = 1$ when:

(a) $|\alpha| < 1$ **(b)** $|\alpha| = 1$ **(c)** $|\alpha| > 1$

98. Sketch the orbit of $f(z) = z^2$ for each of the following values of z:

(a) $z = \cos \frac{2\pi}{3} + i \sin \frac{2\pi}{3}$ **(b)** $z = \cos \frac{\pi}{6} + i \sin \frac{\pi}{6}$
(c) $\cos 17° + i \sin 17°$ **(d)** $\cos 34° + i \sin 34°$

(e) $z = \frac{1}{3} - \frac{1}{3}i$

(f) $z = \sqrt{3} + 3i = 2\sqrt{3}\left(\cos\frac{\pi}{3} + i\sin\frac{\pi}{3}\right)$

(g) $z = \cos 225° + i\sin 225°$

(h) $z = \cos(25\sqrt{10}) + i\sin(25\sqrt{10})$

99. Consider the function $f(z) = z^2$. Describe what happens to the orbit of z if z is:

(a) *inside* the unit circle

(b) *on* the unit circle

(c) *outside* the unit circle

3.6 Julia Sets and the Mandelbrot Set

Julia Sets and the Mandelbrot Set are some of the most intricate and beautiful geometric objects in mathematics. They arise in the study of chaos and dynamical systems. But even without studying these fields, you have the tools to understand how these sets are generated—all it takes is an understanding of the behavior of complex functions and orbits.

This explosion of interest in these sets over the past 20 years is due in part to the fact that their images can easily be generated with a computer. Pictures of the Mandelbrot set now show up routinely in textbooks, on posters, as screensavers, and even in popular culture. What's often masked is the beautiful underlying mathematics used to *generate* the images. This section is an introduction to that process, covering the basics you'll need in order to understand the mathematics behind the pictures. If you want to learn more, you can consult the excellent text [7]. Before we develop the mathematics, let's take a look at some of the pictures. Each of these is generated by a simple algorithm (defined using complex numbers) that we'll develop in this section.

The Mandelbrot Set was named for Benoit Mandelbrot, one of the first to construct the set in 1980. His work began the study of fractal geometry and renewed interest in the study of complex dynamics.

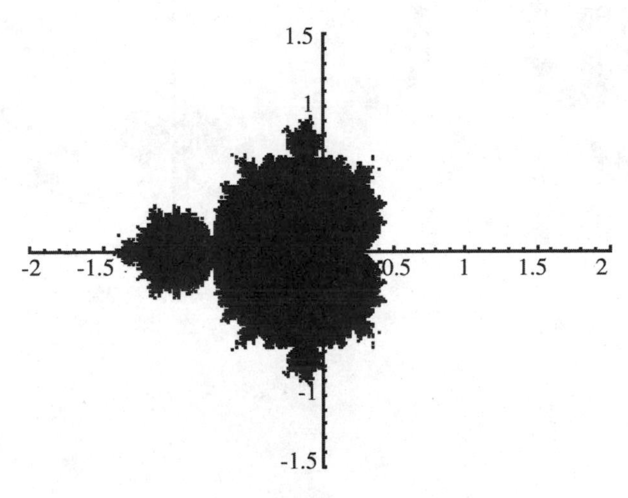

The Mandelbrot Set

filled Julia Set for
$Q(z) = z^2 - 0.26 + .1i$

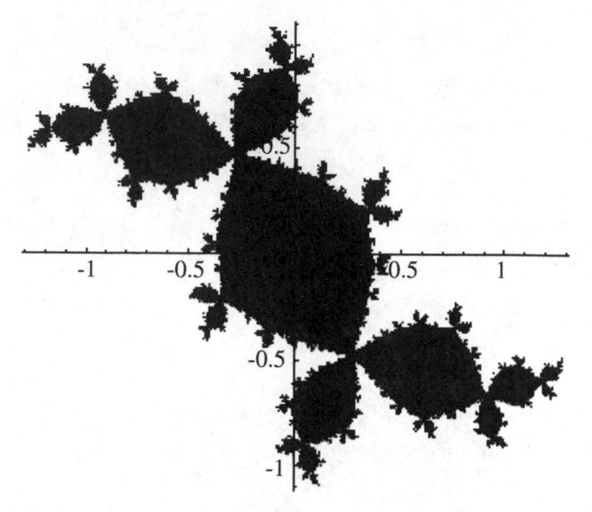

filled Julia Set for
$Q(z) = z^2 + 0.12 + 0.74i$

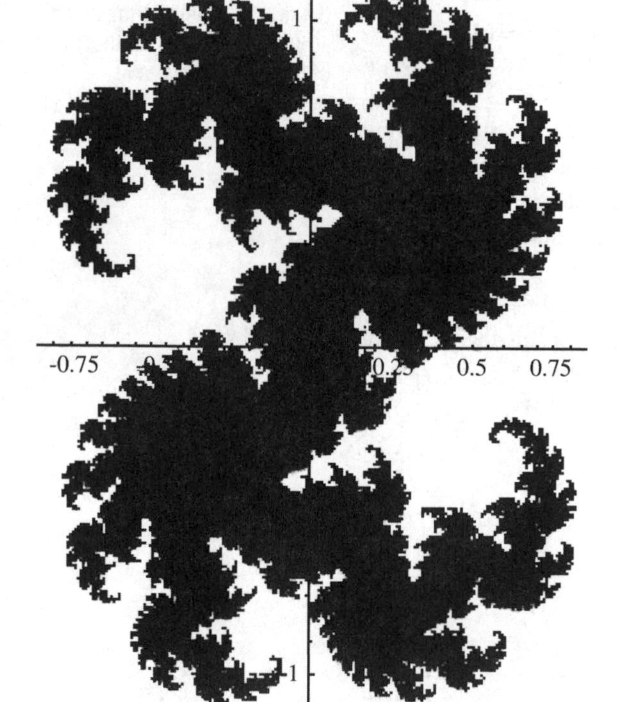

filled Julia Set for
$Q(z) = z^2 + 0.32 + 0.043i$

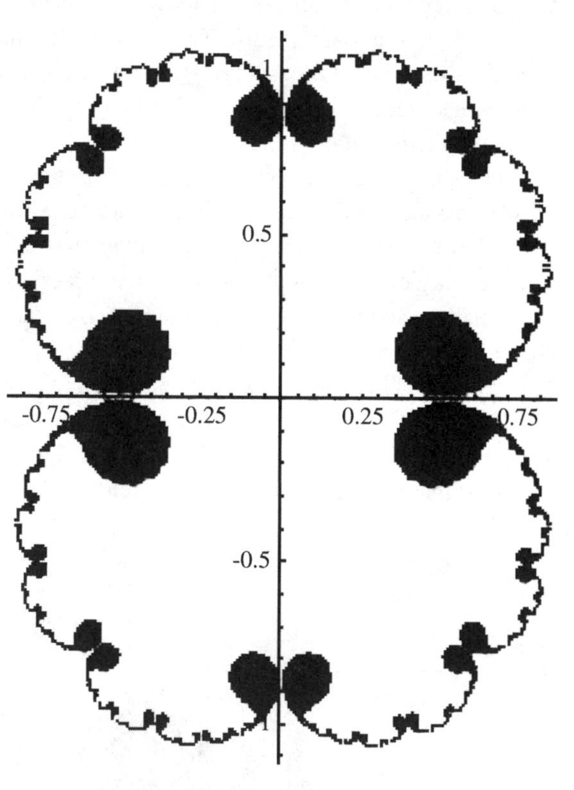

filled Julia Set for
$Q(z) = z^2 + .255$

Historical Perspective

In 1918, when only 25, Gaston Julia published his 199-page masterpiece *Mémoire sur l'iteration des fonctions rationelles.*

As a soldier in the First World War, Julia had been severely wounded. He lost his nose and had to wear a leather strap across his face for the rest of his life. Between several painful operations he carried on his mathematical research in the hospital.

Although he was famous in the 1920s, his work was essentially forgotten until B. Mandelbrot brought it back to prominence in the 1970s through his fundamental computer experiments.

Julia realized how tremendously intricate these Julia sets could be, but had no computer graphics available to see these sets, and as a consequence, his work stopped in the 1930s.

How to Generate the Pictures

Let's start with Julia sets. Each "filled Julia set" is defined by iterating a quadratic function of the form

$$Q_c(z) = z^2 + c$$

where c and z are complex numbers.

Think of c as a parameter. For each c, you get a function:

$$\begin{aligned} Q_2(z) &= z^2 + 2 \\ Q_{.255}(z) &= z^2 + .255 \\ Q_i(z) &= z^2 + i \\ Q_{1+2i}(z) &= z^2 + (1 + 2i) \end{aligned}$$

And, for each c, you can look at the values of z that stay close to the origin when you iterate Q_c. The set of all these points is what we will call the filled Julia set.

Example: Is 3 in the filled Julia set for $Q_{.255}(z) = z^2 + .255$? If you look at the picture on page 146, it sure doesn't look that way. If this picture is accurate, it looks as if you can put a box around the Julia set that will not contain $3 = 3+0i$. Let's check this out.

Iterate $Q_{.255}(z) = z^2 + .255$ starting at $z = 3$.

$$\begin{aligned} Q_{.255}(3) &= 3^2 + .255 \\ &= 9.255 \\ Q^2_{.255}(3) &= Q_{.255}(9.255) \\ &= (9.255)^2 + .255 \\ &= 85.910025 \end{aligned}$$

$$\begin{aligned} Q^3_{.255}(3) &= Q_{.255}(85.910025) \\ &= (85.910025)^2 + .255 \\ &\approx 7380.7874 \\ Q^4_{.255}(3) &\approx Q_{.255}(7380.7874) \\ &= (7380.7874)^2 + .255 \\ &\approx 54476022.8326 \\ &\vdots \end{aligned}$$

From here on out, the outputs will just get bigger and bigger. So we can say that the orbit of 3 tends to infinity, and we can write this as $Q^n_{.255}(3) \to \infty$. Since the orbit of 3 marches off to ∞, 3 is not in the filled Julia set of $Q_{.255}$.

Let's try $0.5 + 0.1i$. It looks (judging from the picture) that this has a chance. Iterate $Q_{.255}(z) = z^2 + i$ starting at $0.5 + 0.1i$.

$$\begin{aligned} Q_{.255}(0.5+0.1i) &= (0.5+0.1i)^2 + 0.255 \\ &= 0.495 + 0.1i \\ Q^2_{.255}(0.5+0.1i) &= Q_{.255}(0.495+0.1i) \\ &= (0.495+0.1i)^2 + 0.255 \\ &\approx 0.490025 + 0.99i \\ Q^3_{0.255}(0.5+0.1i) &\approx Q_{.255}(0.490025+0.99i) \\ &= (0.490025+0.99i)^2 + 0.255 \\ &\approx 0.485323 + 0.09702i \\ Q^4_{.255}(0.5+0.1i) &\approx Q_{.255}(0.485323+0.09702i) \\ &= (0.485323+0.09702i)^2 + 0.255 \\ &\approx 0.48113 + 0.09418i \\ &\vdots \end{aligned}$$

It seems that the iterates of $0.5 + 0.1i$ will never get very big in absolute value. If that's true (and it is), $0.5 + 0.1i$ is in the filled Julia set for $Q_{.255}$.

So when you iterate a function Q_c, the orbits of some points will tend to infinity, and the orbits of other points will not. Orbits that do not tend to infinity are called **bounded**. In a bounded orbit, the absolute value of the output at any stage will be less than some bound. Let's piece all this together and make a formal definition of Julia sets.

If you plot all the points in the filled Julia set, you get pictures like those on page 146. Julia sets are much harder to draw, because they are the boundaries of very complicated sets.

Definition The orbit of z under Q_c is **bounded** if there exists a K such that

$$|Q^m_c(z)| < K$$

for all n.

The **filled Julia set** (denoted by J_c) for Q_c is the set of all complex numbers whose orbits are bounded. The **Julia set** is the boundary of the filled Julia set.

So, to see if a point is in the filled Julia set for Q_c, you have to keep iterating it to see if the orbit marches off to infinity. The next theorem tells you that if the orbit gets far enough away from the origin, you can stop and be certain that the "seed" is not in the filled Julia set.

Theorem 9. *Let $K = \max\{|c|, 2\}$ (that is, K is the greater of the two numbers $|c|$ and 2). If $|Q_c^k(z)| > K$ for some iterate k, then $Q_c^m(z) \to \infty$. Otherwise the orbit is bounded.*

Ways to think about it

Here's a way to picture this fact: Take the bigger number between $|c|$ and 2. Draw a circle of that radius around the origin. If the orbit of z under Q_c ever falls outside of that circle, then we know it will tend to infinity. If it always stays inside the circle, then it is bounded.

If you find an iterate that's outside the circle, you can stop. But if you suspect that all the iterates are inside the circle, you have to prove it.

The proof of Theorem 9 is similar to many proofs in analysis: it involves playing with delicate estimates and properties of absolute value until you get things to work. Indeed, that's probably how Theorem 9 was conceived in the first place. Rather than trying to recreate the ideas that went into the proof, let's look at a more or less standard proof of the theorem, as it might appear in a typical mathematics text. Two properties of absolute value that are used in the proof (and throughout analysis) are the results of problems 28 and 30 on page 111:

- $|zw| = |z||w|$ and, if $n \in \mathbb{Z}^+$, $|z^n| = |z|^n$ (the length of the product is the product of the lengths).
- $|z+w| \leq |z|+|w|$ and $|z-w| \geq |z|-|w|$ (the triangle inequality for complex numbers).

We'll start with a lemma that does most of the work:

Lemma 2. *Suppose $|w| > K$. Then $|Q_c^m(w)| \to \infty$.*

Proof Suppose $|w| > K$. Then $|w| > 2$ and $|w| > c$. Because $|w| > 2$, $|w| - 1 > 1$. That is, $|w| - 1 = 1 + \epsilon$ for some number $\epsilon > 0$.

So, we can argue like this:

$$\begin{aligned}
|Q_c(w)| &= |w^2 + c| \\
&\geq |w|^2 - |c| \quad \text{(see problems 28 and 30 on page 111)} \\
&> |w|^2 - |w| \qquad \text{(because } |w| > |c|\text{)} \\
&= (|w| - 1)|w| \\
&= (1+\epsilon)|w| \qquad \text{(because } |w| - 1 = 1 + \epsilon\text{)}.
\end{aligned}$$

Hence

$$|Q_c(w)| > (1+\epsilon)|w| > |w| > K.$$

Now do it again: Since $|Q_c(w)| > K$, we can apply the same reasoning to $Q_c(w)$:

$$\begin{aligned}
|Q_c^2(w)| &= Q_c(Q_c(w)) \\
&= |Q_c(w)^2 + c| \\
&\geq |Q_c(w)|^2 - |c| \\
&> |Q_c(w)|^2 - |Q_c(w)| \qquad \text{(because } |Q_c(w)| > |c|) \\
&= (|Q_c(w)| - 1)|Q_c(w)| \\
&> (1+\epsilon)((1+\epsilon)|w|) = (1+\epsilon)^2|w|
\end{aligned}$$

$$(|Q_c(w)| - 1 > |w| - 1 = 1 + \epsilon).$$

Inductively, we see that

$$|Q_c^m(w)| > (1+\epsilon)^m|w|.$$

Since $(1+\epsilon)^m \to \infty$, we see that $Q_c^m(w) \to \infty$, as claimed. ■

That's all we need to prove the theorem.

Proof of Theorem 9. Suppose $|Q_c^k(z)| > K$ for some iterate k. We'll show that $Q_c^n(z) \to \infty$. Let $w = Q_c^k(z)$. Then, by Lemma 2, for every positive integer m,

$$|Q_c^m(w)| > (1+\epsilon)^m|w|.$$

We also need to show that if $Q_C^n(z) \to \infty$, then $|Q_C^k(z)| > K$ for some iterate k. We'll leave that as an exercise (problem 102).

So,

$$|Q_c^{(m+k)}(z)| > (1+\epsilon)^m|Q_c^k(z)| > (1+\epsilon)^m \cdot 2 \to \infty.$$

So the orbit marches off to ∞. ■

So, how can you find the filled Julia set for Q_c? Well, for each point z in the plane:

- Compute some number of iterates—say 10,000 of them—of $Q_c(z)$.
- If $|Q_C^n(z)| < \max\{|c|, 2\}$ for all of the iterates, color the point black. It is probably in the filled Julia set. Otherwise, leave the point white.

Of course, you don't have to check every point in the plane. Let $K = \max\{|c|, 2\}$. If a point w is outside the circle of radius K, then it's orbit will certainly tend to infinity. So you only need to check the points inside that circle.

It's no wonder that the study of complex dynamics exploded with the computer revolution!

Still, this tedious algorithm is much better suited to a computer than a person. Notice that no matter how many iterates you check, you can never be *sure* that somewhere down the line the orbit doesn't skip out of the boundary circle (unless you fall onto a fixed point or cycle, of course). Except for the boundary points, you can get a pretty good picture of the filled Julia set using this algorithm.

Example: Find several points that are inside the filled Julia set for Q_i and several points that are outside of it.

Let's try $z = 0$. Iterate $Q_i(z) = z^2 + i$ starting at $z = 0$.

$$\begin{aligned} Q_i(0) &= 0^2 + i \\ &= i \\ Q_i^2(0) &= Q_i(i) \\ &= (i)^2 + i \\ &= -1 + i \\ Q_i^3(i) &= Q_i(-1 + i) \\ &= (-1 + i)^2 + i \\ &= -i \\ Q_i^4(i) &= Q_i(-i) \\ &= (-i)^2 + i \\ &= -1 + i \\ &\vdots \end{aligned}$$

From here on out, the outputs cycle back and forth between $-1 + i$ and $-i$. So, we know that the orbit for 0 ends on the cycle between $-i$ and $-1 + i$, so these three points are all inside the filled Julia set. So is i, since it lands on the same cycle.

Any point z where $|z| > 2$ will be outside the filled Julia set. The point $z = 1$ is also outside the filled Julia set. Notice:

$$\begin{aligned} Q_i(1) &= 1^2 + i \\ &= 1 + i \\ Q_i^2(1) &= Q_i(1 + i) \\ &= (1 + i)^2 + i \\ &= 3i \end{aligned}$$

Since $|3i| = 3 > 2$, we know that the orbit of 1 will tend to infinity. You can test individual complex numbers like this to get a feel for the points that escape and those that do not. This is especially easy if your calculator handles complex numbers (iteration can be done through the **Ans** button). But it takes a computer program to generate enough points to really see the picture. The beautiful filigree of J_i looks like this:

Notice that 0 is in the filled Julia set for Q_i, but 1 is not in the set.

Remember: J_i is the filled Julia set for Q_i.

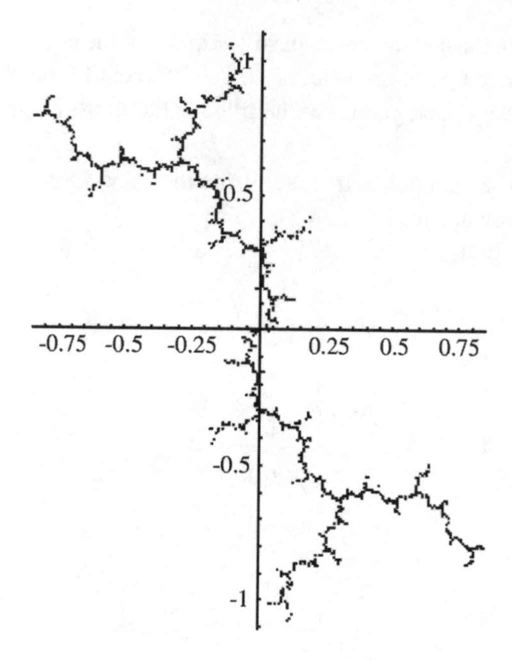

J_i, the filled Julia Set for Q_i

Problem 26 says that z and $-z$ are symmetric with respect to the origin.

The proof shows more than we need. All we needed to show is that the orbits of z and $-z$ are either both bounded or not. The proof shows that they are the *same*. A teacher we know refers to this as the "generosity of mathematics."

But you *don't* need a computer to prove things about what you see. For example, notice that J_i seems to be symmetric with respect to the origin. We can prove that using problem 26 on page 111. In fact, this symmetry holds for any Julia set:

Theorem 10. *Let $c \in \mathbb{C}$. The filled Julia set J_c is symmetric with respect to the origin.*

Proof We will show that $z \in J_c \Leftrightarrow -z \in J_c$. We have

$$Q_c(z) = z^2 + c = (-z)^2 + c = Q_c(-z).$$

So, inductively,

$$Q_c^n(z) = Q_c^n(-z)$$

for every integer n. Hence the orbit of z and the orbit of $-z$ are exactly the same. The result follows. ∎

The Mandelbrot Set

The Mandelbrot set is defined in terms of the Julia sets. Here's how:

The filled Julia sets are different for each function Q_c. For each $c \in \mathbb{C}$, we could test 0 to see if it is in J_c. If it is, we say that c is in the Mandelbrot set.

Definition The **Mandelbrot set** is the set of all the points c where the orbit of 0 under Q_c is bounded. That is, the Mandelbrot set is the set of points c such that 0 is in the filled Julia set for Q_c.

So, when thinking of the Mandelbrot set, each point c in the complex plane represents a *function*:

$$Q_c = z^2 + c.$$

This function has a filled Julia set (all the points whose orbits are bounded). The Mandelbrot set is defined by the following test: If 0 is in the filled Julia set—if the orbit of 0 is bounded—then c is in the Mandelbrot set. If 0 is not in the filled Julia set, then c is not in the Mandelbrot set.

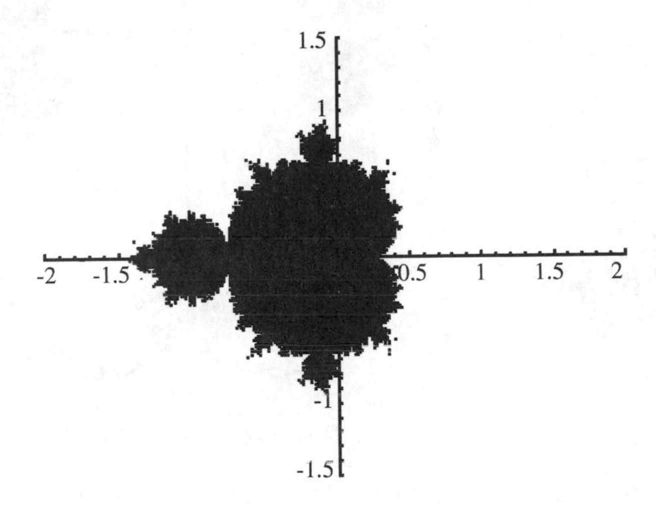

You may have seen color pictures of the Mandelbrot set with this black shape in the middle and colors radiating out. Colors are often used to show how quickly the orbits tend to infinity, and to make the picture more attractive. But the points in the Mandelbrot set itself are not usually shown in color.

Example: Find two points in the Mandelbrot set and two points that are not.

Let $c = 0$. We have the function $Q_0 = z^2$. $Q_0^n(0) = 0$ for every n, so the orbit of 0 is the fixed point 0. So $c = 0$ is in the Mandelbrot set. Similarly, for $c = i$ we have seen that the orbit of 0 under Q_i lands on a cycle of two points, so the point $c = i$ is also in the Mandelbrot set.

If $c = 2$, then

$$Q_2(0) = 2, \quad \text{and} \quad Q_2^2(0) = 2^2 + 2 = 6.$$

Clearly the orbit of 0 will tend to infinity, so $c = 2$ is not in the Mandelbrot set.

If $c = 1.5i$, then we can compute the first few iterates of $Q_{1.5i}(z) = z^2 + 1.5i$, starting with $z = 0$:

$$Q_{1.5i}(0) = 1.5i, \quad Q_{1.5i}^2(0) = -2.25 + 1.5i.$$

Note that $|Q_{1.5i}^2(0)| > 2$, so the orbit of 0 will tend to infinity. So $c = 1.5i$ is not in the Mandelbrot set.

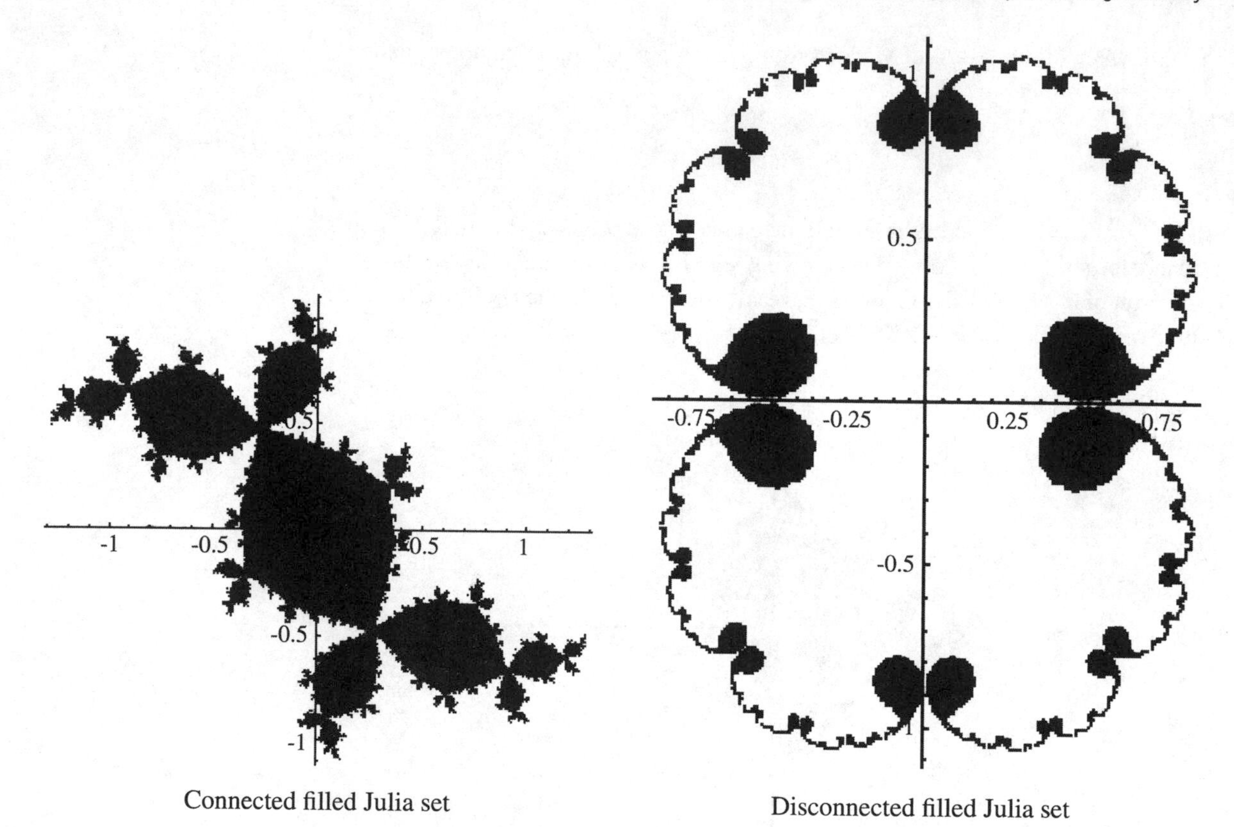

Connected filled Julia set

Disconnected filled Julia set

So the points in the Mandelbrot set have nice, connected Julia sets corresponding to them. In fact, one way to define the Mandelbrot set is as the set of all c-values such that the filled Julia set for Q_c is connected.

For more on Julia sets, the Mandelbrot set, and the connection between the two, visit the Mandelbrot Set explorer on the Web at `http://math.bu.edu/DYSYS/explorer/`.

Problems

100. Iterate $Q_{-1}(z)$ starting with the following seeds. Describe the behavior of the orbit. Do you think it tends to infinity or remains bounded?

(a) $z = 0$ **(b)** $z = i$ **(c)** $z = 0.2 + 0.5i$

101. Iterate $Q_{1-i}(z)$ starting with the following seeds. Describe the behavior of the orbit. Do you think it tends to infinity or remains bounded?

(a) $z = 0$ **(b)** $z = i$

102. Suppose $c \in \mathbb{C}$ and let $K = \max\{|c|, 2\}$. If $Q_c^n(z) \to \infty$, show that $|Q_c^k(z)| > K$ for some iterate k.

103. Consider the function $Q_0 = z^2$.

(a) Find five points that are inside the filled Julia set for Q_0 and five points that are outside the set.

(b) Sketch the filled Julia set for this function.

(c) What is the Julia set? That is, describe the boundary of the filled Julia set.

104. Decide which of the following points are inside the Mandelbrot set and which are outside of it:

(a) $c = -2$ **(b)** $c = -i$ **(c)** $c = 1$

(d) $c = -1$ **(e)** $c = 0.3 - 0.4i$ **(f)** $c = -0.75 + 0.25i$

105. Notice that the Mandelbrot set seems to be symmetric to the real axis. Prove it.

Hint: See problem 27 on page 111 and show that

$$Q_{\overline{c}}(0) = \overline{Q_c(0)}.$$

Notes for Selected Problems

Notes for problem 17 on page 107 If the general proof becomes too messy, try it in a case that's general in principle. Suppose, for example that $f(x) = ax^3 + bx^2 + cx + d$ and $g(x) = rx^2 + sx + t$. Then $\overline{f}(x) = \overline{a}x^3 + \overline{b}x^2 + \overline{c}x + \overline{d}$ and $\overline{g}(x) = \overline{r}x^2 + \overline{s}x + \overline{t}$. Now write out and simplify both sides of each alleged identity and show that you get the same thing on each side.

Notes for problem 44 on page 119 We want to show that $z\overline{z} = 1$. Since z is on the unit circle, you can write $z = \cos\theta + i\sin\theta$ for some angle θ. Then $\overline{z} = \cos\theta - i\sin\theta$. Using these representations, multiply z and $\overline{z}$ together and use the Pythagorean identity.

Notes for problem 50 on page 119 Before we use DeMoivre, note that you can solve this equation by factoring:

$$x^6 - 1 = (x^3 - 1)(x^3 + 1) = (x - 1)(x^2 + x + 1)(x + 1)(x^2 - x + 1).$$

This is only one way to factor it. At least two others have come up in our classes. Can you find them?

So, $x = \pm 1$ and four other numbers that can be found with the quadratic formula applied to each quadratic factor.

But, in general, it isn't easy to factor $x^n - 1$ into linear and quadratic factors, so let's look for another method. We're looking for six numbers whose sixth powers are all 1. Well, if $z^6 = 1$, $|z| = 1$ (problem 29 on page 111), so z is on the unit circle: $z = \cos\theta + i\sin\theta$ for some θ. And we also know that

$$1 = \cos 360^\circ + i\sin 360^\circ.$$

So, we have

$$(\cos\theta + i\sin\theta)^6 = \cos(6\theta) + i\sin(6\theta) = \cos 360^\circ + i\sin 360^\circ.$$

One way to make this work is to set $6\theta = 360^\circ$, so $\theta = 60^\circ$. This produces one root:

$$\zeta = \cos 60^\circ + i\sin 60^\circ = \frac{1}{2} + i\frac{\sqrt{3}}{2}.$$

Another root is clearly 1 (because $1^6 = 1$). We could find other roots by taking other angles for θ, but we can look at it a little more abstractly and get a bonus:

Another way to make this work is to set $6\theta = 720^\circ$, so $\theta = 120^\circ$.

Since $\zeta^6 = 1$, any power of ζ also has a sixth power equal to 1, because

$$(\zeta^e)^6 = (\zeta^6)^e = 1^e = 1.$$

So, now we have plenty of roots. In fact,we have too many; it looks as if this whole sequence of complex numbers satisfies $x^6 - 1 = 0$:

$$1, \zeta, \zeta^2, \zeta^3, \zeta^4, \dots$$

And we know that $x^6 - 1 = 0$ has exactly six roots. Well, the first six of these are distinct, and after that, they repeat. For example,

$$\zeta^8 = \zeta^6\zeta^2 = 1 \cdot \zeta^2 = \zeta^2.$$

Much like the powers of i, these things cycle every 6 (instead of every 4) powers:

$$\begin{aligned}
1 &= \cos 0^\circ + i \sin 0^\circ \\
\zeta &= \cos 60^\circ + i \sin 60^\circ = \frac{1}{2} + i\frac{\sqrt{3}}{2} \\
\zeta^2 &= \cos 120^\circ + i \sin 120^\circ = -\frac{1}{2} + i\frac{\sqrt{3}}{2} \\
\zeta^3 &= \cos 180^\circ + i \sin 180^\circ = -1 \\
\zeta^4 &= \cos 240^\circ + i \sin 240^\circ = -\frac{1}{2} - i\frac{\sqrt{3}}{2} \\
\zeta^5 &= \cos 300^\circ + i \sin 300^\circ = \frac{1}{2} - i\frac{\sqrt{3}}{2} \\
\zeta^6 &= \cos 360^\circ + i \sin 360^\circ = 1 \\
\zeta^7 &= \zeta \\
&\vdots
\end{aligned}$$

If you plot these on the complex plane, you get a regular hexagon. It's this connection between the roots of $x^n - 1 = 0$ and regular n-gons that allowed Gauss to settle the centuries-old question of which regular polygons could be constructed with only straightedge and compass. See [20] for more details.

Notes for problem 70 on page 134 We use Theorem 8. More precisely, since $18 = \frac{90}{5}$ and $54 = \frac{3 \cdot 90}{5}$, we know that $\cos 18^\circ$ and $\cos 54^\circ$ are among the zeros of

$$t_5(x) = 5x - 20x^3 + 16x^5.$$

You could ask a CAS to find the zeros for you (both *Mathematica* and the TI-89 can do it), but that would spoil the fun. Notice that

$$5x - 20x^3 + 16x^5 = x(5 - 20x^2 + 16x^4).$$

So, one root is 0. The remaining factor has degree 4, but it's a quadratic in x^2:

$$5 - 20(x^2) + 16(x^2)^2.$$

So, you can solve for x^2 by the quadratic formula:

$$x^2 = \frac{20 \pm \sqrt{400 - 320}}{32} = \frac{5 \pm \sqrt{5}}{8}$$

so

$$x = \pm \frac{1}{2}\sqrt{\frac{5 \pm \sqrt{5}}{2}}.$$

There are four numbers here. Together with 0, they form (by Theorem 8) the five roots of $t_5(x) = 0$:

$$\cos 18°, \cos 54°, \cos 90° = 0, \cos 126°, \text{ and } \cos 162°.$$

We'll leave it to you to figure out which is which.

The fact that these numbers can be expressed with only square roots (no cube roots, for example) can be used to show that you can construct a regular pentagon with only straightedge and compass.

Notes for problem 72 on page 134 There are many ways to get this identity (many precalculus books have one). We'll sketch an idea that will give you several identities like this at once. It's based on a pair of algebraic identities that come up often enough to be noticed:

$$A = \frac{1}{2}(A + B) + \frac{1}{2}(A - B)$$
$$B = \frac{1}{2}(A + B) + \frac{1}{2}(B - A)$$

Simple enough. Now use these with the multiplication rule for complex numbers:

$$\cos A + i \sin A = \left(\cos \tfrac{1}{2}(A + B) + i \sin \tfrac{1}{2}(A + B)\right)\left(\cos \tfrac{1}{2}(A - B) + i \sin \tfrac{1}{2}(A - B)\right)$$
$$\cos B + i \sin B = \left(\cos \tfrac{1}{2}(A + B) + i \sin \tfrac{1}{2}(A + B)\right)\left(\cos \tfrac{1}{2}(B - A) + i \sin \tfrac{1}{2}(B - A)\right)$$

Multiply out the right-hand sides and compare real parts. You'll end up with expressions for $\cos A$ and $\cos B$. Subtract them, and you'll get the desired identity. By adding them, or by looking at the imaginary parts, you'll get a stash of this type of identity (see problem 74).

In here, you may need to use the fact that cosine is even and sine is odd. For example, $\cos(B - A) = \cos(A - B)$ and $\sin(B - A) = -\sin(A - B)$.

Notes for problem 75 on page 134 You can do this the old-fashioned way (as in precalculus), or you can try this: Write each identity in the form $f(S, C) = 0$ where $S = \sin x$, $C = \cos x$ and f is a polynomial in S and C. Now use problem 41 on page 68 of Chapter 2.

Notes for problem 80 on page 135 Let θ be any angle whatsoever. Then

$$t_k\left(t_\ell(\cos\theta)\right) = t_k\left(\cos(\ell\theta)\right) = \cos(k\ell\theta).$$

The last equality is by problem 79. Similarly,

$$t_\ell\left(t_k(\cos\theta)\right) = \cos(\ell k\theta).$$

So, the polynomials $f(x) = t_k\left(t_\ell(x)\right)$ and $g(x) = t_\ell\left(t_k(x)\right)$ agree for infinitely many inputs (namely, $\cos\theta$ where θ is any angle). So, they are equal as polynomials.

This is a general strategy when proving identities among Chebyshev polynomials: show that both sides are equal at $\cos\theta$ for every angle θ.

Notes for problem 83 on page 135 Just to get a feel for things, let's try it for $k = 3$. The right-hand side expands like this:

$$\sum_{j=0}^{\infty}(-1)^j\binom{3}{2j}x^{3-2j}(1-x^2)^j = \binom{3}{0}x^{3-0}(1-x^2)^0 - \binom{3}{2}x^{3-2}(1-x^2)^1 + \binom{3}{4}x^{3-4}(1-x^2)^2 - \cdots$$

Only the first two terms are nonzero. So, we get

$$x^3 - 3x(1-x^2) = 4x^3 - 3x,$$

which is $t_3(x)$. You might try it for $k = 4$, 5, and 6.

And in general? When you see binomial coefficients, think binomial theorem. And when you see t_k, think $\cos k\theta$. And when you see $\cos k\theta$, think DeMoivre. So, we're thinking "binomial theorem," "Chebyshev," and "DeMoivre" ... Well, here's an idea: To get a formula for $\cos k\theta$, use the fact that

$$\cos k\theta + i\sin k\theta = (\cos\theta + i\sin\theta)^k$$

and expand the right-hand side by the binomial theorem. Here's a start:

$$\begin{aligned}\cos k\theta + i\sin k\theta = (\cos\theta + i\sin\theta)^k &= \binom{k}{0}\cos^k\theta \\ &+ i\binom{k}{1}\cos^{k-1}\theta\sin\theta - \binom{k}{2}\cos^{k-2}\theta\sin^2\theta \\ &- i\binom{k}{3}\cos^{k-3}\theta\sin^3\theta + \binom{k}{4}\cos^{k-4}\theta\sin^4\theta \\ &+ i\binom{k}{5}\cos^{k-5}\theta\sin^5\theta - \cdots\end{aligned}$$

The real part of the right-hand side is $\cos\theta$. So,

$$\cos k\theta = \binom{k}{0}\cos^k\theta - \binom{k}{2}\cos^{k-2}\theta\sin^2\theta + \binom{k}{4}\cos^{k-4}\theta\sin^4\theta - \binom{k}{6}\cos^{k-6}\theta\sin^6\theta\cdots$$

Ah, but $\sin^2\theta = 1 - \cos^2\theta$. So, $\cos k\theta$ can be written as

$$\binom{k}{0}\cos^k\theta - \binom{k}{2}\cos^{k-2}\theta(1-\cos^2\theta) + \binom{k}{4}\cos^{k-4}\theta(1-\cos^2\theta)^2 - \binom{k}{6}\cos^{k-6}\theta(1-\cos^2\theta)^3\cdots$$

This *is* a polynomial, because the $\binom{k}{2j}$ are eventually 0.

But $\cos k\theta$ is also $t_k(\cos\theta)$. So the polynomial $t_k(x)$ and

$$\binom{k}{0}x^k - \binom{k}{2}x^{k-2}(1-x^2) + \binom{k}{4}x^{k-4}(1-x^2)^2 - \binom{k}{6}x^{k-6}(1-x^2)^3\cdots$$

agree for infinitely many inputs (namely $\cos\theta$ where θ is any angle). So they are equal as polynomials.

Neither *Mathematica* nor the TI 89 like the upper bound ∞ in the sum (both systems are usually fine with infinite sums, but here they report a wrong answer (*Mathematica*) or an unsimplified one (TI-89)). But if you stop the sum just before it becomes 0, both systems generate the right polynomials (that is, the Chebyshev polynomials).

In *Mathematica*, one line does it:

```
h[k_] := Expand[
 Sum[Binomial [k,2*j](-1)^j*x^(k-2*j)*(1-x^2)^j,{j,0,Floor[k/2]}]]
```

In the TI-89, it's almost the same:

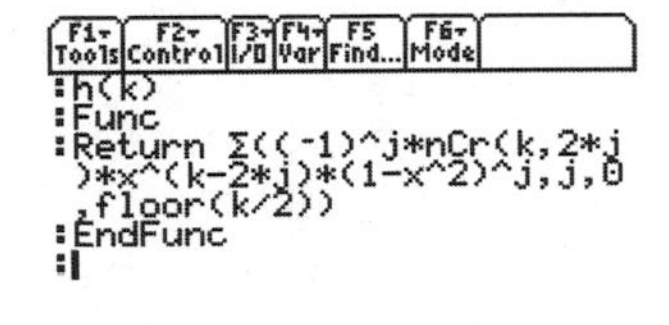


```
h(k)
Func
   Return
     ∑((−1)^j * nCr(k, 2 * j) * x^(k − 2 * j)
          *(1 − x^2)^j, j, 0, floor(k/2))
   EndFunc
```

4

Combinations and Locks

Introduction

Combinatorics, sometimes called "the art of counting without counting" [6], is an active branch of mathematics that has made some inroads into the high school curriculum. Most of these inroads have been in discrete mathematics courses, probability courses, or parts of topics courses. This chapter has some of these kinds of examples—how combinatorics itself can fit into the high school program—but it is also about another use of combinatorial ideas: we want to look at how combinatorics and combinatorial thinking can be used to illuminate ideas from more mainstream high school courses, like algebra.

Combinatorics provides techniques that allow you to figure out how many ways there are to do something without having to make a list of all the ways. For example, if you have three shirts and two pair of jeans, you can make 3×2 outfits. This is an example (albeit an extremely basic one) of a combinatorial result: If you have n objects in set A and m objects in set B, you can make nm ordered pairs whose first "coordinate" comes from A and whose second one comes from B. A more subtle combinatorial result was mentioned in Chapter 1 (page 11):

The set of ordered pairs (a, b) where $a \in A$ and $b \in B$ is suggestively denoted by $A \times B$.

The entries in Pascal's triangle count subsets. Suppose you have a set of five elements, say $\{A, B, C, D, E\}$. How many 3-element subsets are there? There are $\binom{5}{3} = 10$. Here they are:

$$\{A, B, C\}, \{A, B, D\}, \{A, B, E\}, \{A, C, D\}, \{A, C, E\},$$
$$\{A, D, E\}, \{B, C, D\}, \{B, C, E\}, \{B, D, E\}, \{C, D, E\}$$

You can list all the 3-element subsets of a 5-element set with very little trouble. But you'd be hard pressed to list all the 3-element subsets of a 100-element set. Even so, the result says there are $\binom{100}{3} = 161700$ of them.

Another use of combinatorics is the so-called "combinatorial proof." A typical combinatorial proof establishes some identity by showing that both sides represent different ways to perform the same counting process. For example, suppose we take the definition of $\binom{n}{k}$ to be the one above: $\binom{n}{k}$ is the number of k-element subsets of an n-element set. Suppose further that we want to show that these "subset counter" $\binom{n}{k}$ satisfy the same recurrence as the "number pattern" definition of Pascal's triangle (the rows end in 1 and any interior number is the sum of the two above it). That is, using the subset interpretation of $\binom{n}{k}$, we want to show that:

$$\binom{n}{k} = \begin{cases} 1 & \text{if } k = 0 \text{ or if } k = n \text{ (each row starts and ends with 1),} \\ \binom{n-1}{k-1} + \binom{n-1}{k} & \text{if } 0 < k < n \text{ (any interior number is the sum of the two above it).} \end{cases}$$

We could argue like this:

- $\binom{n}{0} = 1$ because there's only one subset of an n-element set that has 0 elements (namely, the empty set).
- $\binom{n}{n} = 1$ because there's only one subset of an n-element set that has n elements (namely, the whole set).
- If $0 < k < n$, we can show that

$$\binom{n}{k} = \binom{n-1}{k-1} + \binom{n-1}{k}$$

This "proof by committee" is an example of what Marvin Freedman, a mathematician who teaches at Boston University, calls a "story proof."

by a "committee" proof: Suppose you have n people and you want to form a k-member committee. The committee is just a set of k people chosen from n, so there are $\binom{n}{k}$ such committees. But let's count the number of committees in a different way. Pick one of the n people, say Nina, and separately count the k-member committees that do and don't contain Nina as a member:

1. **The committees that contain Nina.** Well, there are $k - 1$ slots open, and you have $n - 1$ people who can fill them, so there are

$$\binom{n-1}{k-1}$$

committees that contain Nina.

2. **The committees that do not contain Nina.** Well, there are k slots open, and you have only $n - 1$ people who can fill them (the original n people, but then you exclude Nina), so there are

$$\binom{n-1}{k}$$

committees that don't contain Nina.

But every committee either contains or does not contain Nina, so there are

$$\binom{n-1}{k-1} + \binom{n-1}{k}$$

k-element committees. But there are also $\binom{n}{k}$ such committees. Hence

$$\binom{n}{k} = \binom{n-1}{k-1} + \binom{n-1}{k}$$

as claimed.

Combinatorial proofs often allow you to establish identities (especially identities that involve binomial coefficients) that would be quite difficult and messy using algebra.

In this chapter, we'll look at several applications—of combinatorial results *and* proofs—to some topics connected to high school mathematics. More precisely,

- In section 4.1, we'll give some combinatorial proofs of some interesting identities.
- In section 4.2, we'll look at several combinatorial solutions to the famous "Simplex lock" problem.
- In section 4.4, we'll return to the problem first introduced in section 1.6 of Chapter 1: How one moves back and forth between the powers of x and the "Mahler basis" for polynomials. We'll see that the numbers that show up in these conversions also show up in one solution to the Simplex lock problem! We'll look at some reasons underneath this "coincidence."

The Simplex lock is a project in [15]. Many of the solutions in section 4.2 came from work students did in high school classes.

In addition to producing some very pretty mathematics, these topics will be used to show how the ideas of combinatorics—incisive methods for shortcutting the counting process and combinatorial proofs—can be applied to topics from the high school curriculum.

Problems

1. Suppose you take the "factorial way" to define binomial coefficients. That is, your definition is

$$\binom{n}{k} = \frac{n!}{k!(n-k)!}.$$

Prove that, using this definition, $\binom{n}{k}$ is the number of k-element subsets of an n-element set.

2. Using the factorial way to define binomial coefficients, give a non-combinatorial proof of the fact that, if $0 < k < n$,

$$\binom{n}{k} = \binom{n-1}{k-1} + \binom{n-1}{k}.$$

Get ready to do some algebra.

3. Give a "committee" proof of the identity ($0 \le k \le n$):

$$\binom{n}{k} = \binom{n}{n-k}$$

If you have 20 people, picking a committee of 5 is the same as picking a non-committee of 15.

Pick *two* distinguished people, say, Nina and Nona.

4. Give a committee proof of the identity ($2 \le k \le n-2$):

$$\binom{n}{k} = \binom{n-2}{k-2} + 2\binom{n-2}{k-1} + \binom{n-2}{k}$$

How many 8-element subsets of a 6-element set are there?

5. Suppose you use the "subset way" of defining $\binom{n}{k}$. What would be some sensible ways of defining $\binom{6}{8}$, or in general, $\binom{n}{k}$ when $k > n$?

6. Come up with a reasonable definition of $\binom{-5}{3}$, using one of the definitions from the discussion starting on page 11 of Chapter 1.

It might help to "left justify" the triangle, the way you'd see it in a spreadsheet.

7. Using your results from problems 5 and 6, write out an "expanded" Pascal's triangle that has negative rows and 0s to the right of the rightmost 1. State and explain some patterns in this expanded array.

8. Let $c_{m,k}$ stand for the number of ways you can break a set of n things into k disjoint nonempty subsets. For example, $c_{3,2} = 3$, because there are three ways to break a set of three things into two nonempty subsets. Here they are:

$$\{A, B\} \cup \{C\}$$
$$\{A, C\} \cup \{B\}$$
$$\{C, B\} \cup \{A\}$$

What's analogous to the "every number is the sum of the two above it" rule?

Calculate a few rows of a $c_{m,k}$ triangle, "Pascal triangle style" and describe several ways to characterize the entries.

$\left\langle {m \atop k} \right\rangle$ is the number of *ordered* partitions of an n-element set that have k parts.

9. Let $\left\langle {m \atop k} \right\rangle$ stand for the number of ways you can break a set of m things into k disjoint nonempty subsets, where *the order in which the subsets are listed matters*. For example, $\left\langle {3 \atop 2} \right\rangle = 6$, because there are six ways to break a set of three things into 2-subset sequences. Here they are:

$$\{A, B\} \cup \{C\}$$
$$\{A, C\} \cup \{B\}$$
$$\{C, B\} \cup \{A\}$$
$$\{C\} \cup \{A, B\}$$
$$\{B\} \cup \{A, C\}$$
$$\{A\} \cup \{C, B\}$$

Calculate a few rows of an $\left\langle {m \atop k} \right\rangle$ triangle, "Pascal triangle style" and describe several ways to characterize the entries.

10. Ms. D'Amato likes to take a different route to work every day. She will quit her job the day she has to repeat her route. Her home and work are pictured in the grid of streets below. If she never backtracks (she only travels north or east), how many days will she work at this job?

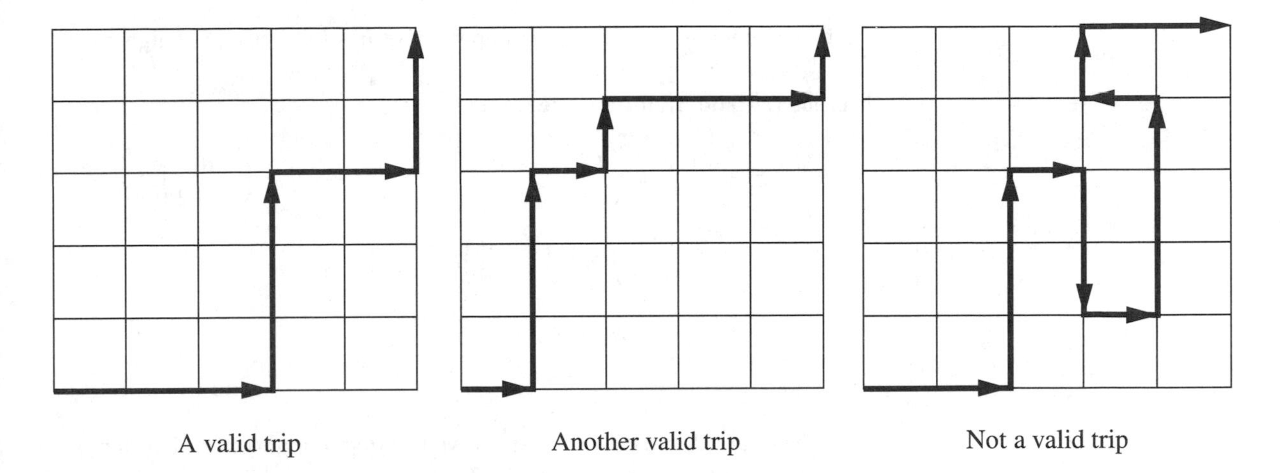

A valid trip

Another valid trip

Not a valid trip

11. How many functions are there from a 5-element set to a 7-element set? From an m-element set to an n-element set?

12. A function is called "one-to-one" if no two elements in the domain end up at the same element of the range. How many one-to-one functions are there from a 5-element set to a 7-element set? From a 7-element set to a 5-element set? From an n-element set to an m-element set?

13. How many one-to-one functions are there from a 20-element set to a 365-element set? How many functions from a 20-element set to a 365-element set are *not* one to one?

14. What's the probability that at least two people in a room of 20 have the same birthday? What if there are 25 people in the room? How many people have to be in the room before the probability of two same-day birthdays is more than .5?

Hint: First find the probability that all 20 people have *different* birthdays.

15. A function is called "onto" if every element in the range gets hit by at least one element in the domain. How many onto functions are there from a 7-element set to a 5-element set? From a 5-element set to a 7-element set? From an m-element set to a k-element set?

Hint: How is this related to problem 9?

4.1 Combinatorial Proofs and Identities

In this section, we'll give a few examples of combinatorial ("story") proofs. The heart of the section is a collection of problems for you to try. Combinatorial proofs provide very simple ways to establish complicated results. In a few lines, you realize that a complex algebraic identity is, in fact, quite simple—if you look at it in the right way. The trade-off is that finding the "right way" to look at it is notoriously difficult. The only way we know to develop the skill is to study some examples and then practice, practice, practice.

Let's begin with one of the most important results in high school algebra:

Theorem 1. (The Binomial Theorem)

$$(a+b)^n = \binom{n}{0}a^n + \binom{n}{1}a^{n-1}b + \binom{n}{2}a^{n-2}b^2 + \cdots + \binom{n}{n-2}a^2b^{n-2} + \binom{n}{n-1}ab^{n-1} + \binom{n}{n}b^n = \sum_{k=0}^{n}\binom{n}{k}a^{n-k}b^k.$$

Proof There are dozens of beautiful ways to prove this theorem, but here's a combinatorial proof. Look at $(a+b)^5$. It's really just

$$(a+b)(a+b)(a+b)(a+b)(a+b).$$

Imagine doing the calculation. If you multiplied all this out, you'd get a sum of terms. You get each term by taking a letter from each parentheses and multiplying them together.

Or, it's the number of ways you can pick two things (two "a"s) from five parentheses. That's $\binom{5}{2}$, which is the same as $\binom{5}{3}$.

For example, you could take b from parentheses 1, 2, and 4 and a from parentheses 3 and 5. That would give you an a^2b^3. But you could also get an a^2b^3 by taking b from parentheses 1, 2, and 3 and a from parentheses 4 and 5. The *coefficient* of a^2b^3 will be all the ways you can pick 3 "b" parentheses and 2 "a" parentheses. And *that* is just the number of ways you can pick three things (three "b"s) from five parentheses. It's $\binom{5}{3}$.

So, you can pick no b's and five a's (that term is $\binom{5}{0}a^5b^0$), one b and four a's (that term is $\binom{5}{1}a^4b^1$), ... So,

$$(a+b)^5 = \binom{5}{0}a^5b^0 + \binom{5}{1}a^4b + \binom{5}{2}a^3b^2 + \binom{5}{3}a^2b^3 + \binom{5}{4}ab^4 + \binom{5}{5}a^0b^5 = \sum_{k=0}^{5}\binom{5}{k}a^{5-k}b^k.$$

The same idea applies in general to $(a+b)^n$. Every term will be of the form $a^{n-k}b^k$ (pick k b's and $n-k$ a's). And the *coefficient* of $a^{n-k}b^k$ is the number of ways you can pick k b's from n parentheses—it's just $\binom{n}{k}$. So the expansion will be the sum of all such terms, and

$$(a+b)^n = \sum_{k=0}^{n}\binom{n}{k}a^{n-k}b^k. \qquad \blacksquare$$

This is just the beginning. By replacing a and b with particular values, you get many interesting numerical identities. For example, replace a and b by 1 and you get

Corollary 1.

Notice that this says that the sum of the entries in the nth row of Pascal's triangle is 2^n.

$$2^n = \binom{n}{0} + \binom{n}{1} + \cdots + \binom{n}{n}$$
$$= \sum_{k=0}^{n} \binom{n}{k}.$$

Yes, this is easy to see by replacing a and b by 1 in the binomial theorem, but let's look at a combinatorial proof of Corollary 1:

Ways to think about it

The right-hand side of the identity in the corollary is the sum of the number of

0-element, 1-element, 2-element, ... , n-element

subsets of an n-element set. So, it's the *total* number of subsets of an n-element set. That gives us the inspiration to count the total number of subsets of an n-element set in another way.

Suppose you have a set of n things, say $A = \{a_1, \ldots, a_n\}$. How many subsets does A have? Well, think about how you'd build a subset. Imagine passing over each element and deciding "yes, I want this one in my subset," or "no, I don't want this one in my subset." You want to count every possible combination of strings of length n composed of "yes" and "no."

So, all yes's corresponds to the whole set, all no's corresponds to the empty set, all strings that have two yes's and the rest no's correspond to all the 2-element sets, and so on.

But there are two choices of a_1 (yes or no), two choices for a_2, two choices for a_3, ... , two choices for a_n. In other words there are

$$\underbrace{2 \times 2 \times 2 \times 2 \cdots \times 2}_{n \text{ times}} = 2^n$$

strings of "yes" and "no," so there are 2^n possible subsets. Hence

$$2^n = \binom{n}{0} + \binom{n}{1} + \cdots + \binom{n}{n}$$

because both sides count the total number of subsets of an n-element set.

Many teachers are always on the lookout for interesting identities involving entries in Pascal's triangle. One source of inspiration is to substitute well-chosen numbers into the binomial theorem (think about what you get if you replace a by 1 and b by -1, for example). Combinatorics offers another treasure house of identities: those that come from thinking about different ways to count the same thing. For example, suppose you want to pick five things from a set of twelve. There are $\binom{12}{5}$ ways to do this. But let's make up a different way to count. Suppose you split your 12-element set into two pieces, say a piece A with

We chose 7 and 5 arbitrarily. Any two nonnegative integers that sum to 12 would work.

seven things and a piece B with five things. Now you could pick your five things by taking some from A and some from B. More precisely, you could take:

None from A and five from B: This can be done in

$$\binom{7}{0}\binom{5}{5} \quad \text{ways}$$

One from A and four from B: This can be done in

$$\binom{7}{1}\binom{5}{4} \quad \text{ways}$$

Two from A and three from B: This can be done in

$$\binom{7}{2}\binom{5}{3} \quad \text{ways}$$

Three from A and two from B: This can be done in

$$\binom{7}{3}\binom{5}{2} \quad \text{ways}$$

Four from A and one from B: This can be done in

$$\binom{7}{4}\binom{5}{1} \quad \text{ways}$$

Five from A and none from B: This can be done in

$$\binom{7}{5}\binom{5}{0} \quad \text{ways}$$

So, we have an unexpected identity:

$$\binom{12}{5} = \binom{7}{0}\binom{5}{5} + \binom{7}{1}\binom{5}{4} + \binom{7}{2}\binom{5}{3} + \binom{7}{3}\binom{5}{2} + \binom{7}{4}\binom{5}{1} + \binom{7}{5}\binom{5}{0}$$

Similarly:

$$\binom{12}{5} = \binom{6}{0}\binom{6}{5} + \binom{6}{1}\binom{6}{4} + \binom{6}{2}\binom{6}{3} + \binom{6}{3}\binom{6}{2} + \binom{6}{4}\binom{6}{1} + \binom{6}{5}\binom{6}{0}$$

Once you get in the swing of things, there are many more variations on this theme, so you can invent all kinds of identities (your students will love you for this).

Generalizing these ideas, we have a useful result:

Theorem 2. (Vandermonde's Identity) *If m, n, and r are nonnegative integers,*

$$\binom{m+n}{r} = \sum_{k=0}^{r} \binom{m}{k}\binom{n}{r-k}.$$

Ways to think about it

Combinatorial proofs are wonderful, but they are not the *only* wonderful proofs for identities. There are at least two other methods that deserve mention.

Algebraic proofs: Consider Vandermonde's identity in Theorem 2. One way to prove it is to show that both sides are the coefficient of the same term in an algebraic simplification. Since the coefficient of a term in a polynomial is unique, both sides must be equal. In the Vandermonde case, we start with the fact that

$$(1+x)^{(m+n)} = (1+x)^m(1+x)^n.$$

The idea is to look at the coefficient of x^r on both sides. On the left-hand side, it is, by the binomial theorem, $\binom{m+n}{r}$. On the right-hand side, imagine expanding $(1+x)^m$ and $(1+x)^n$ by the binomial theorem and multiplying the results together. So, the coefficient of x^r on the right-hand side is

$$\sum_{k=0}^{r} \binom{m}{k}\binom{n}{r-k},$$

and the identity follows.

To get an x^r on the right, you'd have to take all products of the coefficients of x^k (from the first expansion) and x^{r-k} (from the second) and add the results. But the coefficient of x^k in $(1+x)^m$ is $\binom{m}{k}$, and the coefficient of x^{r-k} in $(1+x)^n$ is $\binom{n}{r-k}$.

Computerized proofs: A body of research starting in about the mid-1940s and extending to current work—by the authors of [1] and others—has developed a set of algorithms that can be implemented on a computer and that will supply proofs of binomial coefficient identities. Better yet, if you don't know how to simplify a sum like the right-hand side of Vandermonde's identity, the algorithms will *find* a closed form for you or report with certainty that none exists. The wonderful and readable book [1] contains all the details.

Problems

16. Suppose you have a group of twelve people and you want to form a committee of seven people. The seven people want to pick a subcommittee (from their ranks) of four people. In how many ways can all this be done?

Find at least two ways to count the possibilities. What if you picked the subcommittee first and then "grew" it to the full committee?

17. Show that

$$\binom{12}{7}\binom{7}{4} = \binom{12}{4}\binom{8}{3}$$

18. In general, show that

$$\binom{n}{r}\binom{r}{k} = \binom{n}{k}\binom{n-k}{r-k}$$

In problems 18–20, n, r, and k are integers for which the formulas make sense.

Proof #1: It's a special case of problem 18.

19. Give at least two proofs of the identity

$$r\binom{n}{r} = n\binom{n-1}{r-1}$$

20. Give at least two proofs of the identity

$$(r+1)\binom{n}{r+1} = (n-r)\binom{n}{r}.$$

21. Recall from chapter 1 (section 1.3) the combinatorial polynomials that make up the Mahler basis:

$$\binom{x}{r} = \frac{x(x-1)\cdots(x-r+1)}{r!}$$

Give at least two proofs of the identity

$$\frac{x-r}{r+1}\binom{x}{r} = \binom{x}{r+1}$$

I got this problem from my high school teacher, Frank Kelley.

22. Write out the first few powers of 11. Why do their digits give entries in Pascal's triangle? Will they always be entries in Pascal's triangle?

The hockey stick property: If you start at the end of any row and draw a hockey stick along a diagonal as shown, the sum of the entries on the handle of the stick is the entry at the tip of the blade.

23. Pascal's triangle enjoys a sort of "hockey stick" property:

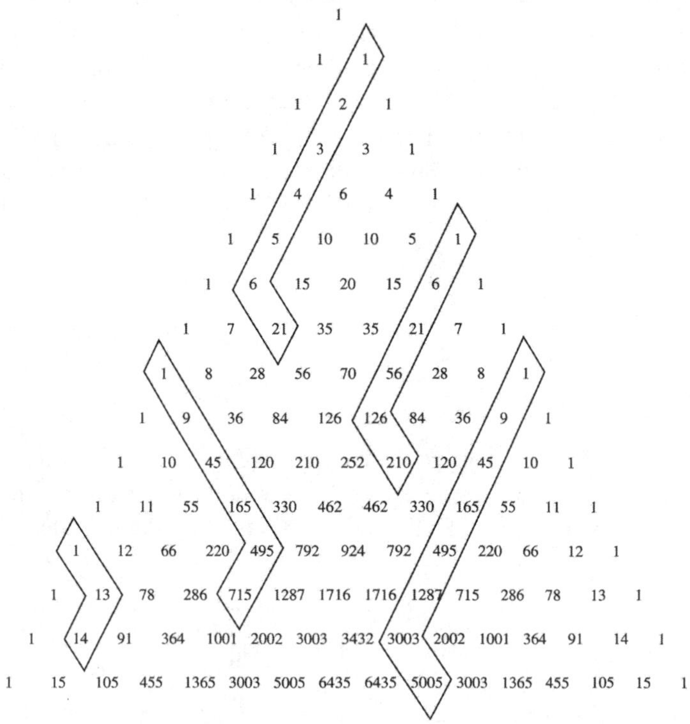

1
1 1
1 2 1
1 3 3 1
1 4 6 4 1
1 5 10 10 5 1
1 6 15 20 15 6 1
1 7 21 35 35 21 7 1
1 8 28 56 70 56 28 8 1
1 9 36 84 126 126 84 36 9 1
1 10 45 120 210 252 210 120 45 10 1
1 11 55 165 330 462 462 330 165 55 11 1
1 12 66 220 495 792 924 792 495 220 66 12 1
1 13 78 286 715 1287 1716 1716 1287 715 286 78 13 1
1 14 91 364 1001 2002 3003 3432 3003 2002 1001 364 91 14 1
1 15 105 455 1365 3003 5005 6435 6435 5005 3003 1365 455 105 15 1

Express the hockey stick property as an identity involving binomial coefficients and prove the identity.

24. Show that

$$\binom{2n}{r} = \sum_{k=0}^{r} \binom{n}{k}\binom{n}{r-k}$$

Hint: Split the $2n$-element set into two equal pieces.

25. Show that

$$\binom{3n}{r} = \sum_{k=0}^{r} \binom{n}{k}\binom{2n}{r-k}$$

26. Show that

$$\binom{m+n}{r+s} = \sum_{k=-r}^{s} \binom{m}{r+k}\binom{n}{s-k}$$

27. Prove Vandermonde's Identity (Theorem 2).

28. Prove that if n is a positive integer,

$$\binom{2n}{n} = \sum_{k=0}^{n} \binom{n}{k}^2$$

The sum of the squares of the entries in the nth row of Pascal's triangle is the middle entry of the $2n$th row? Wow.

29. Prove the following identity:

$$\frac{\binom{m+1}{k}\binom{k+1}{j}}{\binom{m+1}{j}} = \frac{k+1}{m+2-j}\binom{m+2-j}{k+1-j}$$

30. Show that the number of ways you can roll a total of 5 on two dice is the coefficient of x^5 in

$$(x + x^2 + x^3 + x^4 + x^5 + x^6)^2$$

31. What's the number of ways to roll a total of 9 on three dice?

A CAS may help here.

32. What's the sum of *all* the numbers in Pascal's triangle, up to and including the nth row? Prove what you say.

33. Invent three identities involving binomial coefficients by coming up with ways to count the same thing in two different ways.

34. Invent three identities involving binomial coefficients by substituting numbers in the binomial theorem.

35. Give at least two proofs of the fact that

$$\sum_{k=0}^{n} (-1)^k \binom{n}{k} = 0$$

Method #1: Use the binomial theorem.

36. Suppose A is a set with n elements. Show that the number of subsets of A with an even number of elements is the same as the number of subsets of A with an odd number of elements.

37. Establish the identity

$$(x+a+1)^n - (x+a)^n = nx^{n-1} + \binom{n}{n-2}x^{n-2}\left((a+1)^2 - a^2\right) + \binom{n}{n-3}x^{n-3}\left((a+1)^3 - a^3\right) + \cdots + \binom{n}{0}\left((a+1)^n - a^n\right)$$

Compare with problem 80 on page 82 of Chapter 2.

38. Establish the identity

$$4(x+a)^3 + 6(x+a)^2 + 4(x+a) + 1 = 4x^3 + 6x^2\left((a+1)^2 - a^2\right) + 4x\left((a+1)^3 - a^3\right) + \left((a+1)^4 - a^4\right)$$

We'll meet this sequence again in the next chapter.

39. Consider the sequence of numbers defined by

$$B_m = \begin{cases} 1 & \text{if } m = 0 \\ -\dfrac{1}{m+1}\left(\displaystyle\sum_{k=0}^{m-1}\binom{m+1}{k}B_k\right) & \text{if } m > 0 \end{cases}$$

(a) Calculate B_k for $k = 1, \ldots, 6$.

(b) Consider the polynomial in two variables

$$4(x+a)^3 + 6(x+a)^2 + 4(x+a) + 1$$

If this is expanded and each power of a, a^k, is replaced by B_k, show that everything simplifies to $4x^3$.

The cardinality of a finite set is how many elements it has.

40. Find a formula for the sum of the cardinalities of all the subsets of an n-element set.

41. Give at least two proofs of the identity

$$\sum_{k=0}^{n} k\binom{n}{k} = n2^{n-1}$$

42. You can use rods of integer sizes to build "trains" that all share a common length. A "train of length 5" is a row of rods whose combined length is 5. Here are some examples.

Assume you have rods of every possible integer length available. Can you come up with an *algorithm* that will generate all the trains of length n?

Notice that the 1-2-2 train and the 2-1-2 train contain the same rods but are listed separately. If you use identical rods in a different order, this is a separate train.

1	2	2

2	1	2

1	3	1

1	4

5

(a) How many trains of length 5 are there?

(b) Come up with a formula for the number of trains of length n. Prove that your formula is correct.

(c) How many trains of length n have k "cars?"

4.2 The Simplex Lock

What has come to be known as the *Simplex lock problem* has been a staple in my classes, from elementary algebra to advanced courses, for years. This section takes a somewhat deeper look at the problem, developing some of the (student and teacher) approaches we've seen. More precisely, we'll outline some of these approaches, and leave the fun part (filling in the details) to you.

I got the problem from Brian Harvey, who had already been using it for some time with his computer science students. See problem 49 on page 184 for an example of what one of Brian's students did with the problem.

Let's start with the problem. The Simplex company makes a combination lock that is used in many public buildings. It comes in several versions. Here is one:

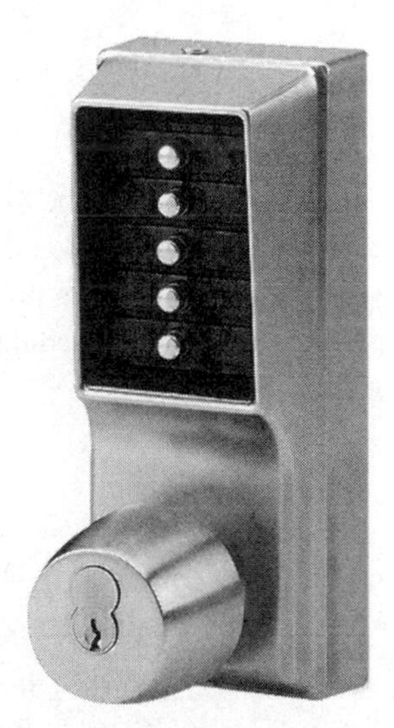

These 5-button devices are purely mechanical (no electronics). You can set the combination using the following rules:

1. A combination is a sequence of 0 or more pushes, each push involving at least one button.
2. Each button may be used at most once (once you press it, it stays in).
3. Each push may include any number of "open" buttons, from one to five.
4. When two or more buttons are pushed at the same time, order doesn't matter.

Notation: $\{\{1,2\},\{3\}\}$ means "press 1 and 2 together, then press 3."

Possible combinations:

- $\{\{1,2\},\{3\}\}$
- $\{\{2,1\},\{3\}\}$
- $\{\{2\},\{1\},\{3\}\}$
- $\{\{1,2,4\},\{3,5\}\}$
- $\{\{1,2,4,3,5\}\}$
- $\{\{1,2\},\{4\},\{3,5\}\}$
- $\{\{3\},\{1,2\}\}$
- $\{\,\}$
- $\{\,2\}$

The company advertises thousands of combinations, and (as we say to our students), the question is, "How many combinations are there? Is the company telling the truth?"

Well, they are (barely), but the value of the project lies less in the answer than in the approaches people take to it.

Almost every time I use this problem with students or teachers, I see an approach that is new to me. It will be fun to compare what you do to the suggestions in the next section.

Problem

Stop here, close the book, and don't open it again until you have thought about the problem for at least two days. Reading ahead will spoil the fun and will hamper your creativity.

4.3 Some Approaches to the Simplex Lock Problem

Here are some approaches to the Simplex lock problem that I've seen over the years.

Enumeration Method Number 1

One strategy is to classify the combinations by the total number of buttons used (0, 1, 2, 3, 4, or 5). Then you can sum these answers.

Suppose, for example, your combination uses only two buttons. Then there are only two possible "shapes" the combination could have:

One student used her solution to the problem of determining all the "rod trains" (problem 42 on page 172) of length 5 to determine all the "shapes."

1	1

press one button then another

2

press two buttons together

For the first shape, there are two pushes. For the first push, there are five possible buttons. After choosing the button used for the first push, there are four buttons left to choose from for the second push. So, there are $5 \times 4 = 20$ possible combinations for the first kind of combination.

For the second shape, there's only one push, and it contains two buttons. There are five buttons to pick from, so there are

$$\binom{5}{2} = 10$$

possible combinations of this type.

So, you can represent the two-button situation this way.

$\boxed{\binom{5}{1}}\boxed{\binom{4}{1}} = 20$

press one button then another

$\boxed{\quad\binom{5}{2}\quad} = 10$

press two buttons together

A total of 30 combinations use two buttons.

Where do the numbers come from? For the first shape, you have five buttons and you have to pick one for the first push. There are $\binom{5}{1}$ ways to do this. Then you only have four buttons left, and you have to pick one for the second push. There are $\binom{4}{1}$ ways to do *this*. So, there are

$$\binom{5}{1}\binom{4}{1} = 20$$

ways to get a combination pressing one button then another.

Using this scheme, the three button situation (which has four possible shapes) would look like this:

Look at the first shape in the second row, for example (press two buttons together, then one more). You have $\binom{5}{2}$ choices for the first push, Then you have to pick one button from the remaining three for the second push. That can be done in $\binom{3}{1}$ ways. So, there are

$$\binom{5}{2}\binom{3}{1} = 30$$

ways to press two buttons and then one more.

$\boxed{\binom{5}{1}}\boxed{\binom{4}{1}}\boxed{\binom{3}{1}} = 60$

three single-button presses

$\boxed{\binom{5}{1}}\boxed{\quad\binom{4}{2}\quad} = 30$

press one button, then two together

$\boxed{\quad\binom{5}{2}\quad}\boxed{\binom{3}{1}} = 30$

press two buttons together, then one more

$\boxed{\quad\binom{5}{3}\quad} = 10$

press three buttons together

A total of 130 presses use three buttons.

The idea is to do this counting scheme for every possible button total from 0 to 5. Here's a summary of what happens:

1. **No buttons**: One combination (the door is always open)
2. **One button**: Five possible combinations ($\binom{5}{1}$).

3. **Two buttons**: We did this above. Two shapes and 30 combinations.
4. **Three buttons**. We did this above. Four shapes and 130 combinations.

We'll leave the remaining two possibilities to you.

Enumeration Method Number 2

There's another enumeration technique that's a little more abstract. That's good, because abstraction lets you look at a bigger picture. This method will lead to an algorithm that lets you figure the number of combinations on a lock with *any* number of buttons.

These numbers are actually the right ones, as we'll see in problem 45.

The basic idea is to consider locks with different numbers of buttons, and then to count the number of combinations that use *all* the buttons on a given lock. So, we imagine a function L (for lock) so that $L(m)$ is the number of combinations on an m-button lock that use all the buttons. We then try to calculate $L(1)$, $L(2)$, $L(3)$, $L(4)$, and $L(5)$. How will this help? Well, suppose someone gave you the following table.

n = Number of buttons on the lock	$L(n)$ = number of combinations that use *all* the buttons
0	1
1	1
2	3
3	13
4	75
5	541

Then we could figure out the total number of combinations on a five-button lock by reasoning like this:

- A combination has to use 0, 1, 2, 3, 4, or 5 buttons.
- There's one combination that uses no buttons.
- To get the number of combinations that use one button, count the number of ways to pick a button (there are $\binom{5}{1} = 5$ ways to do this). On each of these buttons, there is exactly one combination. So, there are $\binom{5}{1} \times 1$ combinations that use exactly one button.
- To get the number of combinations that use two buttons, count the number of ways to pick two buttons to actually use (there are $\binom{5}{2} = 10$ ways to do this). On each of these pairs of buttons, there are exactly three combinations (from the table). So, there are $\binom{5}{2} \times 3 = 30$ combinations that use exactly two buttons.

- To get the number of combinations that use three buttons, count the number of ways to pick three buttons to actually use (there are $\binom{5}{3} = 10$ ways to do this). On each of these triples of buttons, there are exactly 13 combinations (from the table). So, there are $\binom{5}{3} \times 13 = 130$ combinations that use exactly three buttons.
- To get the number of combinations that use four buttons, count the number of ways to pick four buttons to actually use (there are $\binom{5}{4} = 5$ ways to do this). On each of these quadruples of buttons, there are exactly 75 combinations (from the table). So, there are $\binom{5}{4} \times 75 = 375$ combinations that use exactly three buttons.
- To get the number of combinations that use five buttons, count the number of ways to pick five buttons to actually use (there are $\binom{5}{5} = 1$ way to do this). On each of these quintuples of buttons, there are exactly 541 combinations (from the table). So, there are $\binom{5}{5} \times 541 = 541$ combinations that use exactly three buttons.
- Add these up:

Number of buttons used	number of combinations
0	1
1	$\binom{5}{1} \times 1 = 5$
2	$\binom{5}{2} \times 3 = 30$
3	$\binom{5}{3} \times 13 = 130$
4	$\binom{5}{4} \times 75 = 375$
5	$\binom{5}{5} \times 541 = 541$
total	1082

Whoops—there's the answer: 1082. But, as we said earlier, the thrill is in the chase.

So, the total number of combinations for a 5-button lock is:

$$\binom{5}{0}L(0) + \binom{5}{1}L(1) + \binom{5}{2}L(2) + \binom{5}{3}L(3) + \binom{5}{4}L(4) + \binom{5}{5}L(5)$$

Ways to think about it

It remains to show that the table on page 176 is correct. That's the object of problem 45. One way to think about this is to count the combinations that use all the buttons on an m-button lock by the number of pushes used. Here's one way to think about it for $m = 5$.

(*continued*)

The 20 above the first shape in the oval with the 150 tag came from

$$\binom{5}{1}\binom{4}{1}\binom{3}{3}$$

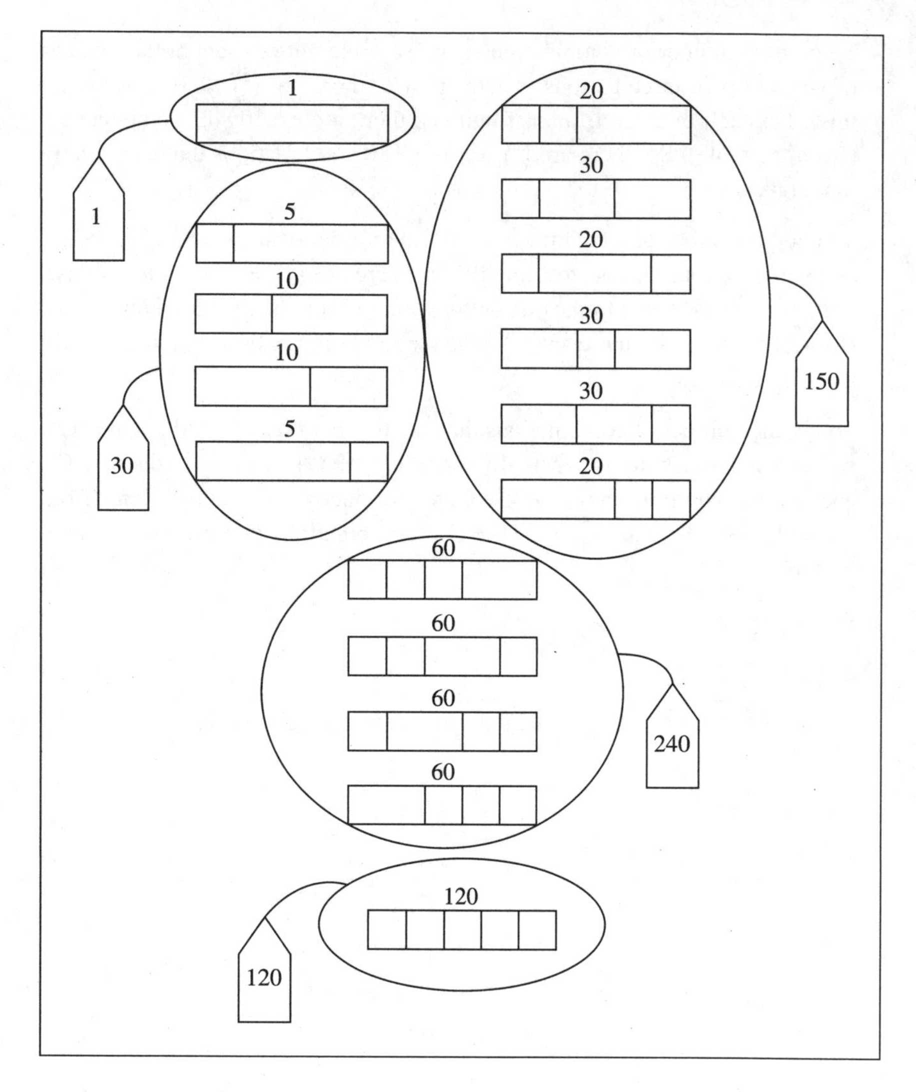

$\left\langle {m \atop k} \right\rangle$ is the number of *ordered* partitions of an m-element set that have k parts.

Notice that the number of combinations on an m-button lock that use all the m-buttons and that contain exactly k pushes is none other than the number we denoted by $\left\langle {m \atop k} \right\rangle$ in problem 9 on page 164. Then the discussion above can be summarized in a table:

k	$\left\langle {5 \atop k} \right\rangle$
0	0
1	1
2	30
3	150
4	240
5	120

Notice how $L(5)$ is exactly half the total number of combinations on a 5-button lock. Could that be a coincidence?

and

$$L(5) = \left\langle {5 \atop 0} \right\rangle + \left\langle {5 \atop 1} \right\rangle + \left\langle {5 \atop 2} \right\rangle + \left\langle {5 \atop 3} \right\rangle + \left\langle {5 \atop 4} \right\rangle + \left\langle {5 \atop 5} \right\rangle = 541$$

You could work out this scheme for any number of buttons. We did it for 0–5 and found

Pushes → Buttons ↓	0	1	2	3	4	5
0	1	0	0	0	0	0
1	0	1	0	0	0	0
2	0	1	2	0	0	0
3	0	1	6	6	0	0
4	0	1	14	36	24	0
5	0	1	30	150	240	120

These are lovely numbers, and the table contains many patterns that may or may not hold in general.

Ways to think about it

The table above is (ignoring the 0's) an "$\left\langle {m \atop k} \right\rangle$ triangle." One way to think about generating the entries in the $\left\langle {m \atop k} \right\rangle$ triangle is to mimic what was done in the diagram on page 177 for five buttons. Another is to think about how you'd build a row from the previous row. For example, knowing the fifth row, how could you calculate, say, $\left\langle {6 \atop 4} \right\rangle$. Well, $\left\langle {6 \atop 4} \right\rangle$ counts the number of combinations on a 6-button lock that uses four pushes. Suppose the buttons are numbered 1–6, and concentrate on button number 6. Button 6 could show up in a push with other buttons or as a push all by itself.

- Suppose it is in a push with other buttons. Then if you "forget the 6," you have a combination on 1–5 with 4 pushes.
- Suppose it is a push all by itself. Then if you "forget the 6," you have a combination on 1–5 with 3 pushes.

Now imagine working from five buttons to six. You could get a 4-push 6-button combination from looking at the combinations that use 1–5 and adding in the number 6 button, either in an existing push or as a push all by itself.

- There are $\left\langle {5 \atop 4} \right\rangle$ combinations on 1–5 that use four pushes, and you could put the 6 in any one of those pushes. Hence there are $4\left\langle {5 \atop 4} \right\rangle = 960$ combinations on six buttons for which 6 is in an existing push.

And . . .

(continued)

This method generalizes to an approach to problem 47 on page 183.

- There are $\binom{5}{3}$ combinations on 1–5 that use three pushes, and you could put 6 in its own push at the start of the combination, between any two pushes, or at the end of the combination. You can check that for a combination with three pushes, there are four "holes" in which you can put an extra push—at the front, between the first two pushes, between the second two pushes, or at the end. Hence there are $4\binom{5}{3} = 600$ combinations on six buttons for which 6 is in an existing push.

The punchline is that

$$\left\langle {6 \atop 4} \right\rangle = 4\left\langle {5 \atop 4} \right\rangle + 4\left\langle {5 \atop 3} \right\rangle = 1560$$

Combinatorics

People who have worked with the previous method and similar ones have abstracted some patterns in the calculations that have allowed them to deal with the problem in more generality (any number of buttons) and at a more abstract level (without all the arithmetic with specific numbers and all the calculations with the various shapes).

Giving names to things often helps with the abstraction: You name something, then, making believe it's a real thing, you look for some of its properties. For example, let

$T(m)$ = the number of combinations on an m-button lock, and remember that

$L(m)$ = the number of combinations on an m-button lock that uses all m buttons.

Think of $T(m)$ and $L(m)$ as unknowns that you (desperately) want to find. Since there's two of them, you'll need two equations relating them. If you've mucked around with special cases, you may have noticed this:

Theorem 3.

$$T(m) = \sum_{k=0}^{m} \binom{m}{k} L(k).$$

A combinatorial proof again.

Proof For each k between 0 and m, here's a two-step way to count all the combinations that use exactly k buttons:

1. Pick k buttons to actually use in the combination. This can be done in $\binom{m}{k}$ ways.
2. Once you pick the buttons to use, count all the combinations that use all k buttons you have picked. This can be done in $L(k)$ ways.

So, there are $\binom{m}{0}L(0)$ combinations that use no buttons, $\binom{m}{1}L(1)$ that use one button, $\binom{m}{2}L(2)$ that use two buttons, and so on. Add them up and you get the total number of combinations that use *any* number of buttons. That's what the theorem claims:

$$T(m) = \sum_{k=0}^{m} \binom{m}{k} L(k). \qquad \blacksquare$$

That's one equation. For the second, notice that for a five-button lock, the number of combinations that use all five buttons is 541, exactly half the total number of combinations. If you play with other numbers of buttons, this seems to be the case in general. In other words

Theorem 4.

$$T(m) = 2L(m).$$

Another way to think of this result is that the number of combinations that use all the buttons is the same as the number of combinations that use fewer than all the buttons.

Proof A common way to show that two sets have the same size is to set up a one-to-one correspondence between them. That is, if you have one set, P, and another set, R, and you want to show that they have the same number of elements, set up a function that sends the elements of P to the elements of R so that every element of P is assigned to exactly one element of R and all the elements of R are hit.

Imagine P is the set of combinations that use all m buttons, and let R be the set of combinations that use fewer than m buttons. For each combination in P, you can get a combination in R by "forgetting the last push." For example, if $m = 5$ and you have

$$\{1, 2, 4\}\{5\}\{3\}$$

you assign it to

$$\{1, 2, 4\}\{5\}.$$

Now,

- This assignment will produce every combination that uses fewer than m buttons. Why? Well, take a combination that uses fewer than m buttons. Take all the buttons that it doesn't use and make a new push from those, putting it on the end of what you have. This produces a combination that uses all the buttons and that gets assigned to the one you started with.
- No combination that uses fewer than m buttons gets "hit" twice by this assignment. Why? Well, two combinations that use all m buttons that look the same when you forget the last push must differ only in the last push. But you have no *choice* for the last push of a combination that uses all m buttons: It has to contain all the buttons that aren't in the previous pushes.

It follows (if you think about it) that our assignment is one-to-one, and the theorem is proved. $\blacksquare$

Well, we have two equations:

$$T(m) = 2L(m)$$

and

$$T(m) = \sum_{k=0}^{m} \binom{m}{k} L(k) = \binom{m}{0} L(0) + \binom{m}{1} L(1) + \cdots + \binom{m}{n-1} L(m-1) + \binom{m}{m} L(m).$$

So,

$$2L(m) = \binom{m}{0} L(0) + \binom{m}{1} L(1) + \cdots + \binom{m}{m-1} L(m-1) + \binom{m}{m} L(m).$$

The last term on the right is just $L(m)$; subtract it from both sides and get

$$L(m) = \binom{m}{0} L(0) + \binom{m}{1} L(1) + \cdots + \binom{m}{m-1} L(m-1).$$

What a beautiful recurrence. It will let us calculate the $L(m)$ in terms of previous values of $L(k)$. And you can double each $L(m)$ to get the total number of combinations $T(m)$ on an m-button lock. Look:

$$L(0) = 1$$

$$L(1) = \binom{1}{0} L(0) = 1; \quad T(1) = 2$$

$$L(2) = \binom{2}{0} L(0) + \binom{2}{1} L(1) = 3; \quad T(2) = 6$$

$$L(3) = \binom{3}{0} L(0) + \binom{3}{1} L(1) + \binom{3}{2} L(2) = 13; \quad T(3) = 26$$

$$L(4) = \binom{4}{0} L(0) + \binom{4}{1} L(1) + \binom{4}{2} L(2) + \binom{4}{3} L(3) = 75; \quad T(4) = 150$$

$$L(5) = \binom{5}{0} L(0) + \binom{5}{1} L(1) + \binom{5}{2} L(2) + \binom{5}{3} L(3) + \binom{5}{4} L(4) = 541; \quad T(5) = 1082$$

Let's state all this as a theorem:

Theorem 5. *The number of combinations on an m-button lock ($m > 0$) is $2L(m)$ where*

$$L(0) = 1$$

and

$$L(m) = \sum_{k=0}^{m-1} \binom{m}{k} L(k) \quad \text{if } m > 0.$$

Theorem 5 allows you to quickly calculate the values of L (and hence, of T). A CAS or calculator makes the job even easier. For your enjoyment, here are the first 16 values:

m	$L(m)$	$T(m)$
0	1	1
1	1	2
2	3	6
3	13	26
4	75	150
5	541	1082
6	4683	9366
7	47293	94586
8	545835	1091670
9	7087261	14174522
10	102247563	204495126
11	1622632573	3245265146
12	28091567595	56183135190
13	526858348381	1053716696762
14	10641342970443	21282685940886
15	230283190977853	460566381955706

Approaches to the Simplex lock problem abound. We suggest a couple more in the following problems. See [21] for an interesting paper on the problem.

Notice that the equation $T(m) = 2L(m)$ doesn't hold if $m = 0$. Where does the proof of Theorem 4 break down if $m = 0$? The case $m = 0$ is sometimes called the "pathological case." Why?

Problems

43. Find a method not described in this section for counting the number of combinations on a Simplex lock.

44. Finish the details of enumeration method 1 (page 174).

45. Show that the numbers in the table on page 176 are correct.

46. Show that the numbers in the table on page 179 are correct, and extend the table for three more rows.

47. Prove Theorem 6.

Theorem 6. *The numbers $\left\langle {m \atop k} \right\rangle$ defined on page 178 satisfy the recurrence*

$$\left\langle {m \atop k} \right\rangle = \begin{cases} 1 & \text{if } m = k = 0, \\ 0 & \text{if } k > m \text{ or } k < 0, \\ k\left(\left\langle {m-1 \atop k-1} \right\rangle + \left\langle {m-1 \atop k} \right\rangle\right) & \text{if } 0 \le k \le m. \end{cases}$$

That is, $\left\langle {m \atop k} \right\rangle$ is the number of combinations on an m-button lock that use all the m-buttons and that contain exactly k pushes.

48. Suppose we define a function γ of two variables, built by a "convolution" of lock numbers and binomial coefficients:

$$\gamma(n,m) = \sum_{k=1}^{m} \genfrac{\langle}{\rangle}{0pt}{}{m}{k} \binom{n}{k}$$

$$= \genfrac{\langle}{\rangle}{0pt}{}{m}{1}\binom{n}{1} + \genfrac{\langle}{\rangle}{0pt}{}{m}{2}\binom{n}{2} + \cdots + \genfrac{\langle}{\rangle}{0pt}{}{m}{m}\binom{n}{m}$$

Investigate the function γ, perhaps by filling in this table:

$m \rightarrow$ $n \downarrow$ $\gamma(n,m)\searrow$	1	2	3	4	5	6
1						
2						
3						
4						
5						
6						

Explain anything that you see.

Brian is the author of a 3-volume set of books [5] that teach serious programming—and a ton of mathematics—primarily to high school students.

49. Brian Harvey, who teaches computer science at Berkeley and who first put me onto the Simplex lock problem in 1986, often teaches summer courses to high school students from the area. He reports on the work of one student from Oakland:

> "One of my high school students just came in with the following idea: You make a picture like Pascal's triangle, except that in adding entries from one row to the next, each number on row N has a weight, which is its position within the row (starting from 1). So
>
> $1_{[1]}$
>
> $1_{[1]}$ $1_{[2]}$
>
> $1_{[1]}$ $3_{[2]}$ $2_{[3]}$
>
> $1_{[1]}$ $7_{[2]}$ $12_{[3]}$ $6_{[4]}$
>
> $1_{[1]}$ $15_{[2]}$ $50_{[3]}$ $60_{[4]}$ $24_{[5]}$
>
> etc. The numbers in brackets are the weights. So for example the 7 in the fourth row is $1 \cdot 1 + 3 \cdot 2$. The 50 in the fifth row is $7 \cdot 2 + 12 \cdot 3$. And the sum of each row is (starting at $n = 0$) the number of combinations in an n-button Simplex lock."

Notice that the sum of the entries in row n is $T(n)$, not $L(n)$. Brian's student arrived at his method by looking for patterns. The justification came later.

Investigate this method. If $P(n,k)$ stands for the element in the nth row and kth column (Pascal's triangle style), express the student's recurrence in symbols. Does the method work? Do the numbers in each row have any "lock-theoretic" significance?

50. Here's an unusual method for generating $L(n)$. See if you can come up with an argument for why it works:

Consider the operation Λ on polynomials where

$$\Lambda(f)(x) = (x+1)f(x+1) + xf(x)$$

So, for example,

$$\begin{aligned}\Lambda(2x+1) &= (x+1)\big(2(x+1)+1\big) + x(2x+1) \\ &= 4x^2 + 6x + 3\end{aligned}$$

Next, consider the sequence of polynomials $s(n)$ generated by the following rule:

$$s(n) = \begin{cases} 1 & \text{if } n = 0 \\ \Lambda(s(n-1)) & \text{if } n > 0 \end{cases}$$

A CAS quickly generates some of the $s(n)$:

n	$s(n)$
0	1
1	$1+2x$
2	$3+6x+4x^2$
3	$13+30x+24x^2+8x^3$
4	$75+190x+180x^2+80x^3+16x^4$
5	$541+1470x+1560x^2+840x^3+240x^4+32x^5$
6	$4683+13454x+15540x^2+9520x^3+3360x^4+672x^5+64x^6$

Well, look at that: The constant term of $s(n)$ seems to be $L(n)$—the number of combinations on an n-button lock that use all the buttons. Is it true? If so, why? If not, when does it break down?

Λ is a creature similar to Δ:

$$\Delta(f)(x) = f(x+1) - f(x)$$

and

$$\Lambda(f)(x) = (x+1)f(x+1) + xf(x)$$

Why would anyone ever think up an operation like Λ? Or the polynomials $s(n)$?

Find and explain some interesting patterns in the sequence of polynomials. What happens if you evaluate each polynomial at 1? To see what's going on, it really helps to generate a few by hand.

4.4 Connections to the Mahler Basis

This section was motivated by a seeming coincidence. In Chapters 1 and 2, we investigated ways to represent polynomials in different "bases." The usual way to express a polynomial is as a linear combination of powers of x. For example, suppose

$$f(x) = \frac{x^5}{40} - \frac{x^4}{3} + \frac{53x^3}{24} - \frac{14x^2}{3} - \frac{37x}{30} + 7.$$

For some purposes (Newton's difference formula, for example), it turns out to be more useful to express f in terms of the Mahler basis

$$f(x) = 3\binom{x}{5} - 2\binom{x}{4} + 5\binom{x}{3} - 4\binom{x}{1} + 7\binom{x}{0}.$$

This is problem 66 on page 37 of Chapter 1.

These Mahler polynomials take values of binomial coefficients at nonnegative integers. When expanded, they look like this:

k	$\binom{x}{k}$	Expanded version
0	1	1
1	x	x
2	$\dfrac{x(x-1)}{2!}$	$\dfrac{-x+x^2}{2}$
3	$\dfrac{x(x-1)(x-2)}{3!}$	$\dfrac{2x-3x^2+x^3}{6}$
4	$\dfrac{x(x-1)(x-2)(x-3)}{4!}$	$\dfrac{-6x+11x^2-6x^3+x^4}{24}$
5	$\dfrac{x(x-1)(x-2)(x-3)(x-4)}{5!}$	$\dfrac{24x-50x^2+35x^3-10x^4+x^5}{120}$
6	$\dfrac{x(x-1)(x-2)(x-3)(x-4)(x-5)}{6!}$	$\dfrac{-120x+274x^2-225x^3+85x^4-15x^5+x^6}{720}$

Problem 63 produced a recursive method for generating the $a_{m,k}$.

The problems in section 1.6 of Chapter 1 looked at the conversion formulas between "normal" and Mahler bases. In particular, problem 61 showed that

$$x^m = \sum_{k=0}^{m} a_{m,k}\binom{x}{k}$$

where

$$a_{m,k} = k^m - \binom{k}{1}(k-1)^m + \binom{k}{2}(k-2)^m - \cdots + (-1)^{k-1}\binom{k}{k-1}1^m.$$

... and 0^0 is taken to be 1.

We can calculate the coefficients $a_{m,k}$ and put them in a matrix (here, up to x^5):

$x^m \downarrow$ $\quad \binom{x}{k} \rightarrow$ $a_{m,k} \searrow$	$\binom{x}{0}$	$\binom{x}{1}$	$\binom{x}{2}$	$\binom{x}{3}$	$\binom{x}{4}$	$\binom{x}{5}$
x^0	1	0	0	0	0	0
x^1	0	1	0	0	0	0
x^2	0	1	2	0	0	0
x^3	0	1	6	6	0	0
x^4	0	1	14	36	24	0
x^5	0	1	30	150	240	120

To check this, write each $\binom{x}{k}$ as $\frac{x(x-1)\cdots(x-k+1)}{k!}$ and simplify.

So, for example,

$$x^4 = 0\binom{x}{0} + 1\binom{x}{1} + 14\binom{x}{2} + 36\binom{x}{3} + 24\binom{x}{4}.$$

Oh my. Look back at the table on page 179:

Pushes $(k) \rightarrow$ / Buttons $(m) \downarrow$ $\left\langle {m \atop k} \right\rangle \searrow$	0	1	2	3	4	5
0	1	0	0	0	0	0
1	0	1	0	0	0	0
2	0	1	2	0	0	0
3	0	1	6	6	0	0
4	0	1	14	36	24	0
5	0	1	30	150	240	120

This is a table of the $\left\langle {m \atop k} \right\rangle$, the number of combinations on an m-button lock that use all m buttons and have k-pushes. But it seems to be the same (well, it *is* the same up to 5) as the table for the $a_{m,k}$, the conversion coefficients that express powers of x as a Mahler expansion!

Maybe it *isn't* always the same—mathematics is full of examples of tables that match for a long time but then disagree. We even saw how to produce such false matches in Chapter 1. One needs to be careful not to jump to conclusions. But, in this case, we have a genuine theorem, and a surprising one at that. Angelo DiDomenico, a retired (after over 40 years) high school teacher in Massachusetts says, "Mathematics is generous. It gives you more than you ask for."

Why should this be so? One reason is buried in the problem sets. Problem 63 on page 36 of Chapter 1 showed that the $a_{m,k}$ are determined by the relations:

$$a_{m,k} = \begin{cases} 1 & \text{if } m = k = 0, \\ 0 & \text{if } k > m \text{ or } k < 0, \\ k(a_{m-1,k-1} + a_{m-1,k}) & \text{if } 0 \le k \le m. \end{cases}$$

And by Theorem 6 on page 183 of this chapter, you showed that the numbers $\left\langle {m \atop k} \right\rangle$ satisfy the recurrence

$$\left\langle {m \atop k} \right\rangle = \begin{cases} 1 & \text{if } m = k = 0, \\ 0 & \text{if } k > m \text{ or } k < 0, \\ k\left(\left\langle {m-1 \atop k-1} \right\rangle + \left\langle {m-1 \atop k} \right\rangle\right) & \text{if } 0 \le k \le m. \end{cases}$$

Since the two functions are defined by the same recursive definition, they are equal for all pairs (m, k) in their domain. The details of the proof are fussy, but essentially, it's a proof by induction on m (see problem 51).

Let's state our result in a slightly different way as a theorem:

Theorem 7. *If m, k are nonnegative integers, then we have a polynomial identity*

$$x^m = \sum_{k=0}^{m} \left\langle {m \atop k} \right\rangle \binom{x}{k}.$$

So, the coefficient of $\binom{x}{k}$ needed to express x^m in terms of the Mahler basis is just the number of combinations on an m-button Simplex lock that use all the buttons and k pushes.

This is just another way to say that $a_{m,k} = \left\langle {m \atop k} \right\rangle$.

Proof We've already outlined one proof that uses induction (problem 51). Let's look at a more combinatorial proof. We start with something you may have established in problem 48 on page 184:

Lemma 1. *For any positive integers n, m,*

$$n^m = \sum_{k=0}^{m} \left\langle {m \atop k} \right\rangle \binom{n}{k}.$$

Proof By problem 11 on page 165, the left-hand side counts the number of functions from an m element set (say T) to an n-element set (say, S).

But to build such a function, you have to pick some subset of elements in S to actually get hit. Say you pick k elements (there are $\binom{n}{k}$ ways to do this). Once you pick a "target" k-element subset of S, you have to pick some elements of T to go to the first element of the target, some more to go to the second element of $S, \ldots$, and the remaining elements go to the kth element of the target. This is the same as building a k-push combination—you have to pick some elements of T to go into the first push, some more to go into the second push, ... , and the remaining elements go into the kth push. So, there are $\left\langle {m \atop k} \right\rangle$ ways to build the function. Hence there are

$$\left\langle {m \atop k} \right\rangle \binom{n}{k}$$

functions that have a k-element target. And therefore there are

$$\sum_{k=0}^{m} \left\langle {m \atop k} \right\rangle \binom{n}{k}$$

functions in all. ■

Now for the proof of Theorem 7, note that the polynomials x^m and

$$\sum_{k=0}^{m} \left\langle {m \atop k} \right\rangle \binom{x}{k}$$

agree for every nonnegative integer n, so, they are equal as polynomials by the "function implies form theorem" (Corollary 3 on page 65 of Chapter 2).

From now on, we'll use both notations—$a_{m,k}$ and $\left\langle {m \atop k} \right\rangle$—interchangeably.

Corollary 2. *For nonnegative m and k,*

$$a_{m,k} = \left\langle {m \atop k} \right\rangle.$$

Proof The polynomials $\sum_{k=1}^{m} a_{m,k} \binom{x}{k}$ and

$$\sum_{k=0}^{m} \left\langle {m \atop k} \right\rangle \binom{x}{k}$$

are both equal to x^m, so they are equal. This implies their "coefficients" are equal (Chapter 1, problem 69 on page 37). ■

Finally, using problem 61 on page 36 of Chapter 1, we have a new way to calculate the lock numbers:

Corollary 3. *For nonnegative m and k,*

We need to take 0^0 to be 1 here.

$$\left\langle {m \atop k} \right\rangle = k^m - \binom{k}{1}(k-1)^m + \binom{k}{2}(k-2)^m - \cdots + (-1)^{k-1}\binom{k}{k-1}1^m.$$

Ways to think about it

So, we have found two explanations for the fact that the same numbers show up in what seem to be completely different contexts. Have we "explained" the surprise? From a mathematical point of view, of course we have. The combinatorial proof, especially, seems to shed light on the mystery: the number of combinations on an m-button lock that use all m buttons and that have k pushes is just the number of functions from an m-element set to a k-element that "hit" each of the k elements at least once.

But, even after you've been through the proofs and have explained away the mystery, doesn't it still seem eerie that these numbers show up in such different situations?

Problems

51. Use the facts that

$$a_{m,k} = \begin{cases} 1 & \text{if } m = k = 0 \\ 0 & \text{if } k > m \text{ or } k < 0 \\ k(a_{m-1,k-1} + a_{m-1,k}) & \text{if } 0 \le k \le m \end{cases}$$

and

$$\left\langle {m \atop k} \right\rangle = \begin{cases} 1 & \text{if } m = k = 0 \\ 0 & \text{if } k > m \text{ or } k < 0 \\ k\left(\left\langle {m-1 \atop k-1} \right\rangle + \left\langle {m-1 \atop k} \right\rangle\right) & \text{if } 0 \le k \le m \end{cases}$$

to give a proof by induction (on m) that $a_{m,k} = \left\langle {m \atop k} \right\rangle$ for all nonnegative integers m and k.

52. Show that the number of functions from an m element set *onto* a k-element set is $\left\langle {m \atop k} \right\rangle$.

Problem 15 on page 165 again. A function is "onto" if it "uses up" its range—every element in the range is the image of something in the domain.

53. Show that the number of functions from an m element set to a k-element set that are *not* onto is:

$$\binom{k}{1}(k-1)^m - \binom{k}{2}(k-2)^m + \cdots + (-1)^k\binom{k}{k-1}1^m$$

If you didn't look at these problems at the time, it would be a good thing to do now.

In Chapter 1, you looked at the problem of "going the other way": converting the Mahler polynomials back to powers of x. In problems 62 and 64 on page 36 of Chapter 1, you showed that if

$$\binom{x}{m} = \frac{x(x-1)(x-2)(x-3)\cdots(x-m+1)}{m!}$$

$$= \sum_{k=0}^{m} b_{m,k}x^k,$$

then the $b_{m,k}$ satisfy the recurrence:

$$b_{m,k} = \begin{cases} 1 & \text{if } m = k = 0 \\ 0 & \text{if } k > m \text{ or } k < 0 \\ \dfrac{b_{m-1,k-1} - (m-1)b_{m-1,k}}{m} & \text{if } 0 \le k \le m \end{cases}$$

Concentrate on the repetition—the rhythm of the calculations.

54. Fill in the next four rows of the $b_{m,k}$ table. Describe and explain any patterns you see:

$\binom{x}{m}\downarrow$ $\quad x^k \rightarrow$ $b_{m,k} \searrow$	x^0	x^1	x^2	x^3	x^4	x^5
$\binom{x}{0} = 1$	1	0	0	0	0	0
$\binom{x}{1} = x$	0	1	0	0	0	0
$\binom{x}{2} = \frac{x(x-1)}{2}$	0	$-\frac{1}{2}$	$\frac{1}{2}$	0	0	0
$\binom{x}{3} = \frac{x(x-1)(x-2)}{6}$	0	$\frac{1}{3}$	$-\frac{1}{2}$	$\frac{1}{6}$	0	0
$\binom{x}{4} = \frac{x(x-1)(x-2)(x-3)}{24}$	0	$-\frac{1}{4}$	$\frac{11}{24}$	$-\frac{1}{4}$	$\frac{1}{24}$	0
$\binom{x}{5} = \frac{x(x-1)(x-2)(x-3)(x-5)}{120}$	0	$\frac{1}{5}$	$-\frac{5}{12}$	$\frac{7}{24}$	$-\frac{1}{12}$	$\frac{1}{120}$

Problems 55–58 look at the relationship between the $a_{m,k}$ and the $b_{m,k}$.

A problem with a point . . .

55. Simplify

$$a_{3,0}(b_{0,0}) + a_{3,1}(b_{1,0} + b_{1,1}x) + a_{3,2}(b_{2,0} + b_{2,1}x + b_{2,2}x^2) + a_{3,3}(b_{3,0} + b_{3,1}x + b_{3,2}x^2 + b_{3,3}x^3)$$

56. Simplify

$$a_{m,0}(b_{0,0}) + a_{m,1}(b_{1,0} + b_{1,1}x) + a_{m,2}(b_{2,0} + b_{2,1}x + b_{2,2}x^2) + \cdots + a_{m,m}(b_{m,0} + b_{m,1}x + \cdots + b_{m,m}x^m)$$

57. Simplify

$$b_{m,0}\left(a_{0,0}\binom{x}{0}\right)+b_{m,1}\left(a_{1,0}\binom{x}{0}+a_{1,1}\binom{x}{1}\right)$$
$$+\,b_{m,2}\left(a_{2,0}\binom{x}{0}+a_{2,1}\binom{x}{1}+a_{2,2}\binom{x}{2}\right)+\cdots$$
$$+\,b_{m,m}\left(a_{m,0}\binom{x}{0}+a_{m,1}\binom{x}{1}+\cdots+a_{m,m}\binom{x}{m}\right)$$

58. Calculate the matrix products

We assume in problem 58 that you know how to multiply matrices. If that's not familiar, all you need is the very basics, and that's something you can look up in any linear algebra book (and in many high school algebra books).

(a)

$$\begin{pmatrix} 1 & 0 & 0 & 0 & 0 & 0 \\ 0 & 1 & 0 & 0 & 0 & 0 \\ 0 & -\frac{1}{2} & \frac{1}{2} & 0 & 0 & 0 \\ 0 & \frac{1}{3} & -\frac{1}{2} & \frac{1}{6} & 0 & 0 \\ 0 & -\frac{1}{4} & \frac{11}{24} & -\frac{1}{4} & \frac{1}{24} & 0 \\ 0 & \frac{1}{5} & -\frac{5}{12} & \frac{7}{24} & -\frac{1}{12} & \frac{1}{120} \end{pmatrix}\begin{pmatrix} 1 & 0 & 0 & 0 & 0 & 0 \\ 0 & 1 & 0 & 0 & 0 & 0 \\ 0 & 1 & 2 & 0 & 0 & 0 \\ 0 & 1 & 6 & 6 & 0 & 0 \\ 0 & 1 & 14 & 36 & 24 & 0 \\ 0 & 1 & 30 & 150 & 240 & 120 \end{pmatrix}$$

(b)

$$\begin{pmatrix} 1 & 0 & 0 & 0 & 0 & 0 \\ 0 & 1 & 0 & 0 & 0 & 0 \\ 0 & 1 & 2 & 0 & 0 & 0 \\ 0 & 1 & 6 & 6 & 0 & 0 \\ 0 & 1 & 14 & 36 & 24 & 0 \\ 0 & 1 & 30 & 150 & 240 & 120 \end{pmatrix}\begin{pmatrix} 1 & 0 & 0 & 0 & 0 & 0 \\ 0 & 1 & 0 & 0 & 0 & 0 \\ 0 & -\frac{1}{2} & \frac{1}{2} & 0 & 0 & 0 \\ 0 & \frac{1}{3} & -\frac{1}{2} & \frac{1}{6} & 0 & 0 \\ 0 & -\frac{1}{4} & \frac{11}{24} & -\frac{1}{4} & \frac{1}{24} & 0 \\ 0 & \frac{1}{5} & -\frac{5}{12} & \frac{7}{24} & -\frac{1}{12} & \frac{1}{120} \end{pmatrix}$$

Will the same phenomenon hold for larger sizes?

The numbers $a_{m,k}$ (or $\left\langle {m \atop k} \right\rangle$, depending on your preference) and $b_{m,k}$ are related to some famous numbers in combinatorics known as the *Stirling numbers*. There are two kinds of Stirling numbers, cleverly called Stirling numbers of the first and second kinds. There are many ways to define them, but here's one that's related to the topic of this section.

See the beautiful book [11] for a detailed account of Stirling numbers and many more topics that are connected to this chapter.

Recall from Chapter 2, section 2.3, page 72, the notion of "falling powers:"

So, $x^{\underline{m}} = m!\binom{x}{m}$.

$$x^{\underline{m}} = x(x-1)(x-2)(x-3)\cdots(x-m+1)$$

One way to think about Stirling numbers is as conversion factors between powers of x and falling powers of x. More precisely:

The alternating sign is separated out to keep the Stirling numbers positive, because they have other combinatorial uses. See problem 66 for an example.

(a) The Stirling numbers of the first kind, $\left|{m \atop k}\right|$, are defined by the equation

$$x^{\underline{m}} = \sum_{k=0}^{m} (-1)^{m-k} \left|{m \atop k}\right| x^k$$

(b) The Stirling numbers of the second kind, $\left\{ {m \atop k} \right\}$, are defined by the equation

$$x^m = \sum_{k=0}^{m} \left\{ {m \atop k} \right\} x^{\underline{k}}$$

59. Generate a few rows of the table for $\left| {m \atop k} \right|$. Find and explain some patterns in the table.

60. Show that

$$(-1)^{m+k} m! b_{m,k} = \left| {m \atop k} \right|$$

61. Show that the Stirling numbers of the first kind satisfy the recurrence

$$\left| {m \atop k} \right| = (m-1) \left| {m-1 \atop k} \right| + \left| {m-1 \atop k-1} \right|$$

62. Generate a few rows of the table for $\left\{ {m \atop k} \right\}$. Find and explain some patterns in the table.

63. Show that

$$\left\langle {m \atop k} \right\rangle = k! \left\{ {m \atop k} \right\}$$

64. Show that the Stirling numbers of the second kind satisfy the recurrence

$$\left\{ {m \atop k} \right\} = k \left\{ {m-1 \atop k} \right\} + \left\{ {m-1 \atop k-1} \right\}$$

See problem 66 if you're stuck.

65. Show that

$$\left\{ {m \atop k} \right\} = c_{m,k}$$

where $c_{m,k}$ is defined in problem 8 on page 164.

66. Imagine a variant on the Simplex lock where the order of the pushes doesn't matter.

(a) How many combinations are there on a five-button lock of this type that use all five buttons and have three pushes?

(b) How many combinations are there on a five-button lock of this type that use all five buttons?

(c) How many combinations are there on a five-button lock of this type?

So, [1, 4, 2][3, 5] is the same as [3, 5][1, 4, 2] but *different* from [1, 2, 4][3, 5]. There'd have to be some sort of signal to show that a push is over, and the lock would just store the collection of pushes to see if it matched the combination.

67. Imagine a variant on the Simplex lock where the order of the pushes doesn't matter, but the order that you press buttons within a push *does* matter.

(a) How many combinations are there on a five-button lock of this type that use all five buttons and have three pushes?

(b) How many combinations are there on a five-button lock of this type that use all five buttons?

(c) How many combinations are there on a five-button lock of this type?

Notes for Selected Problems

Note for problem 6 on page 164 The definition that we will use is

$$\binom{n}{k} = \frac{n(n-1)(n-2)\cdots(n-k+1)}{k!}$$

Thus, we have

$$\binom{-5}{3} = \frac{(-5)(-6)(-7)}{3!} = -35$$

We can also "back-track" up the rows of Pascal's Triangle, using the fact that any interior number is the sum of the two above it. Below is the regular Pascal's Triangle "left justified" and zeros to the right of the right-most 1.

$$\begin{matrix} 1 & 0 & 0 & 0 & 0 \\ 1 & 1 & 0 & 0 & 0 \\ 1 & 2 & 1 & 0 & 0 \\ 1 & 3 & 3 & 1 & 0 \\ 1 & 4 & 6 & 4 & 1 \end{matrix}$$

Now for the (-1)st row. It must begin with a 1, and the sum of any two numbers equals the number below them.

Row −1	1	−1	1	−1	1	...
Row 0	1	0	0	0	0	...
Row 1	1	0	0	0	0	...
Row 2	1	2	1	0	0	...
Row 3	1	3	3	1	0	...
Row 4	1	4	6	4	1	...

Try expanding the triangle in this fashion and you should see that the third number on the (-5)th row equals -35.

Note for problem 10 on page 164 Any route that Ms. D'Amato takes involves five east moves and five north moves. For example, the route below can be represented by "EEENNNEENN."

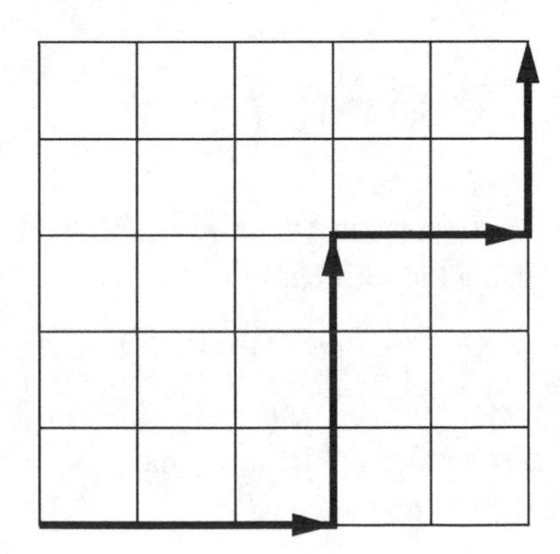

Thus, we need to find the number of 10-letter words using five E's and five N's. This is given by

$$\frac{10!}{5!5!} = 252$$

Note for problem 11 on page 165 Let T be an m-element set and S be an n-element set. Thus,

$$T = \{a_1, a_2, \ldots, a_m\} \quad \text{and} \quad S = \{b_1, b_2, \ldots, b_n\}$$

Suppose f is a function from T to S. There are n possibilities for $f(a_1)$, namely $f(a_1) = b_1$, $f(a_1) = b_2$, ..., or $f(a_1) = b_n$. Likewise, there are n possibilities for $f(a_2)$, n possibilities for $f(a_3)$, and so on. Therefore, there are

$$\underbrace{n \times n \times \cdots \times n}_{m \text{ times}} = n^m$$

possible functions from T to S.

Note for problem 20 on page 170 Suppose you have a group of twelve people and you want to form a committee of eight people. The eight people want to pick a leader of their committee. This can be done in

$$8 \cdot \binom{12}{8} \quad \text{ways.}$$

$\binom{12}{8}$ represents the number of 8-person committees. Multiply it by 8 because any one of the eight members of a committee can be the leader.

We can also pick the seven non-leaders of the committee and choose a leader amongst the five remaining people. This can be done in

$$5 \cdot \binom{12}{7} \quad \text{ways.}$$

Therefore, we have the identity

$$8 \cdot \binom{12}{8} = 5 \cdot \binom{12}{7}$$

It's not hard to see that this can be generalized to the desired formula.

Note for problem 23 on page 170 The hockey stick property can be expressed by

$$\sum_{k=j}^{n-1} \binom{k}{j} = \binom{n}{j+1}$$

where j is a nonnegative integer less than n. This identity can be proved by induction on n and by using the formula

$$\binom{n}{j+1} = \binom{n-1}{j} + \binom{n-1}{j+1}$$

Note for problem 29 on page 171 First, we must assume that $k + 1 - j \geq 0$. Otherwise, $\binom{m+2-j}{k+1-j}$ will have no meaning.

We'll leave it up to you to prove the identity for the special cases $k+1-j=0$ (or $k+1=j$) and $k+1-j=1$ (or $k=j$).

So, now assume that $k+1-j>1$ or $k>j$. Note that

$$\binom{m+1}{k} = \frac{(m+1)(m)(m-1)\cdots(m+2-k)}{k(k-1)(k-2)\cdots 1}$$

$$\binom{k+1}{j} = \frac{(k+1)(k)(k-1)\cdots(k+2-j)}{j!}$$

$$\frac{1}{\binom{m+1}{j}} = \frac{j!}{(m+1)(m)(m-1)\cdots(m+2-j)}$$

Therefore, we have

$$\begin{aligned}\frac{\binom{m+1}{k}\binom{k+1}{j}}{\binom{m+1}{j}} &= \frac{(m+1)(m)(m-1)\cdots(m+2-k)}{(m+1)(m)(m-1)\cdots(m+2-j)}\cdot\frac{(k+1)(k)(k-1)\cdots(k+2-j)}{k(k-1)(k-2)\cdots 1}\\ &= (m+1-j)(m-j)(m-1-j)\cdots(m+2-k)\cdot\frac{(k+1)(k)(k-1)\cdots(k+2-j)}{k(k-1)(k-2)\cdots 1}\\ &= (m+1-j)(m-j)(m-1-j)\cdots(m+2-k)\cdot\frac{k+1}{(k+1-j)(k-j)(k-1-j)\cdots 1}\end{aligned}$$

Now, multiply by $\frac{m+2-j}{m+2-j}$ and rearrange ...

$$\begin{aligned}\frac{\binom{m+1}{k}\binom{k+1}{j}}{\binom{m+1}{j}} &= \frac{k+1}{m+2-j}\cdot\frac{(m+2-j)(m+1-j)\cdots(m+2-k)}{(k+1-j)(k-j)\cdots 1}\\ &= \frac{k+1}{m+2-j}\cdot\frac{(m+2-j)(m+1-j)\cdots\big((m+2-j)-(k+1-j)+1\big)}{(k+1-j)!}\\ &= \frac{k+1}{m+2-j}\cdot\binom{m+2-j}{k+1-j}\end{aligned}$$

Note for problem 30 on page 171 $(x+x^2+x^3+x^4+x^5+x^6)^2$ is the same as

$$(x+x^2+x^3+x^4+x^5+x^6)(x+x^2+x^3+x^4+x^5+x^6)$$

Now, imagine multiplying this out, and think about how you get the term x^5. One way is to multiply x from the first parenthesis and x^4 from the second parenthesis. You can also multiply x^3 from the first parenthesis and x^2 from the second parenthesis. And so on. The coefficient of x^5 will give the number of ways you can pick x^n from the first parenthesis and x^m from the second parenthesis such that $x^n x^m = x^{n+m} = x^5$. Thus, it gives the number of ways you can pick n from the first set and m from the second set such that $n+m=5$.

Note for problem 41 on page 172 The left-hand side of the equation gives the sum of the cardinalities of all the subsets of an n-element set. (See problem 40.)

Here's another way to count this sum. First, list every subset of an n-element set. For example, if $n=3$, the following would be listed:

$$\{\ \}, \{A\}, \{B\}, \{C\}, \{A, B\}, \{A, C\}, \{B, C\}, \{A, B, C\}$$

Now, we just count the total number of letters above. We know that there are

$$\sum_{k=0}^{n} \binom{n}{k} = 2^n$$

subsets in total. And half of them contain the element A (Why?). Thus, there are

$$\frac{1}{2} \cdot 2^n = 2^{n-1}$$

subsets containing A. In other words, there are 2^{n-1} total A's in the list above. Likewise, there are 2^{n-1} total B's and C's. Thus, there are $n \cdot 2^{n-1}$ letters in total.

Note for problem 49 on page 184 The student's recurrence is given by

$$P(n,k) = k \cdot P(n-1,k) + (k-1) \cdot P(n-1,k-1)$$

Using this formula, we generated the fifth row of this triangle.

$$1_{[1]} \qquad 31_{[2]} \qquad 180_{[3]} \qquad 390_{[4]} \qquad 360_{[5]} \qquad 120_{[6]}$$

These numbers do in fact add to 1082, but what does each number represent? Well, here is the table from page 179:

k	$\left\langle {5 \atop k} \right\rangle$
0	0
1	1
2	30
3	150
4	240
5	120

It seems that

$$P(n,k) = \left\langle {n \atop k} \right\rangle + \left\langle {n \atop k-1} \right\rangle$$

For example, the third number of the fifth row, namely 180, can be obtained by adding $\left\langle {5 \atop 3} \right\rangle$ and $\left\langle {5 \atop 2} \right\rangle$. This, along with Theorem 4, explains why the sum of each row of this triangle gives the number of combinations in an n-button Simplex lock.

Finally, to show why

$$P(n,k) = \left\langle {n \atop k} \right\rangle + \left\langle {n \atop k-1} \right\rangle$$

let

$$Q(n,k) = \left\langle {n \atop k} \right\rangle + \left\langle {n \atop k-1} \right\rangle$$

and show that $Q(n, k)$ satisfies the same recurrence as $P(n, k)$. This can be done fairly easily using Theorem 6.

Note for problem 66a on page 192 Consider the combination

$$\{\{1, 2\}, \{4\}, \{3, 5\}\}$$

On a regular Simplex lock, there are 3! different combinations that use these three pushes. They are

$$\begin{array}{c} \{\{1,2\},\{4\},\{3,5\}\} \\ \{\{1,2\},\{3,5\},\{4\}\} \\ \{\{4\},\{1,2\},\{3,5\}\} \\ \{\{4\},\{3,5\},\{1,2\}\} \\ \{\{3,5\},\{1,2\},\{4\}\} \\ \{\{3,5\},\{4\},\{1,2\}\} \end{array}$$

On this variant, however, these six combinations count as one.

In a similar manner, we can group together all simplex lock combinations that use the same three pushes into one. And each group contains 3! or six combinations. Thus, there are

$$\frac{\left\langle {5 \atop 3} \right\rangle}{3!} = \left\{ {5 \atop 3} \right\} = 25 \quad \text{combinations.}$$

By the way, this variant of the Simplex lock is equivalent to the scenario given in problem 8 on page 164. (Why?)

Note for problem 66b on page 192 There are

$$\begin{aligned} &\left\{ {5 \atop 0} \right\} + \left\{ {5 \atop 1} \right\} + \left\{ {5 \atop 2} \right\} + \left\{ {5 \atop 3} \right\} + \left\{ {5 \atop 4} \right\} + \left\{ {5 \atop 5} \right\} \\ &= 0 + 1 + 15 + 25 + 10 + 1 \\ &= 52 \quad \text{combinations in total.} \end{aligned}$$

Note for problem 66c on page 192 Does Theorem 4 also apply to this version of the Simplex lock?

5

Sums of Powers

Introduction

In this last chapter, we come full circle and return to a problem that came up when we were fitting functions to tables in section 1.1 of Chapter 1. We'll make use of much of the mathematics we developed between then and now, as well as some other mathematics you may already know. One goal of this chapter is to develop the formulas for the sums of the first n integers, squares, cubes, and, in general, mth powers.

Remember how this came up. We were working with difference tables like this:

Input	Output	Δ	Δ^2	Δ^3
0	1	−2	14	12
1	−1	12	26	12
2	11	38	38	12
3	49	76	50	12
4	125	126	62	12
5	251	188	74	
6	439	262		
7	701			

By now, we're used to looking across the "0th" row and writing down the Mahler expansion. But another technique was to work from right to left, realizing that each column "accumulates" the values to the right of it.

This is sort of a "spreadsheet" approach.

Input	Output	Δ	Δ^2	Δ^3
0	1	−2	14	12
1	−1	12	26	12
2	11	38	38	12
3	49	76	50	12
4	125	126	62	12
5	251	188	74	
6	439	262		
7	701			

Getting Δ^2 from Δ^3

Input	Output	Δ	Δ^2	Δ^3
0	1	−2	14	12
1	−1	12	26	12
2	11	38	38	12
3	49	76	50	12
4	125	126	62	12
5	251	188	74	
6	439	262		
7	701			

Getting Δ from Δ^2

So, in the Δ^2 column, the output for n is 14 plus the sum of the numbers in the Δ^3 column from 0 to $n-1$, and in the Δ column, the output for n is -2 plus the sum of the numbers in the Δ^2 column from 0 to $n-1$. In general, the output in any column at n is the sum of the outputs in the column to the right, from 0 to $n-1$, plus the output at 0 in the column you're in.

Going from 0 to $n-1$ means that you're adding n things, right?

So a formula for the Δ^2 column is "n goes to 14 plus a bunch of 12s." More precisely, it's

$$n \mapsto 14 + \sum_{k=0}^{n-1} 12 = 14 + 12n.$$

And the Δ column is "n goes to -2 plus a bunch of $(14+12k)$s." More precisely, it's

$$n \mapsto -2 + \sum_{k=0}^{n-1}(14 + 12k).$$

Using the argument from page 6 of Chapter 1, this equation becomes

$$\begin{aligned} n \mapsto -2 + \underbrace{(14 + 14 + \cdots + 14)}_{n \text{ times}} &+ 12(0 + 1 + 2 + \cdots + (n-1)) \\ = -2 + \sum_{k=0}^{n-1} 14 + 12\sum_{k=0}^{n-1} k \\ = -2 + 14n + 12\frac{n(n-1)}{2} = 6n^2 + 8n - 2. \end{aligned}$$

The last sum is evaluated using the formula

$$\sum_{k=0}^{n-1} k = \frac{n(n-1)}{2}.$$

Finally, using this rule for the Δ column, we go from the Δ column to the Output column, we calculate like this:

$$\begin{aligned} n &\mapsto 1 + \sum_{k=0}^{n-1} 6k^2 + 8k - 2 \\ &= 1 + 6\left(\sum_{k=0}^{n-1} k^2\right) + 8\left(\sum_{k=0}^{n-1} k\right) - 2\left(\sum_{k=0}^{n-1} 1\right). \end{aligned}$$

We leave some of the "pencil details" to you.

Ah, look at these sums. The rightmost one is just n. The middle one is (once again) the sum of the integers from 0 to $n-1$, our old friend $n(n-1)/2$. And the leftmost one is the sum of the *squares* of the integers from 0 to $n-1$. If only we had a formula for it.

Note that, except for the rightmost sum, starting at 0 is the same as starting at 1. We include 0 for consistency.

Well, there is such a formula (it's $n(n-1)(2n-1)/6$), so a little algebra allows us to get a formula for the output column. But do you see a method in the wings? Equipped with formulas for the sums of powers (and maybe a CAS), you could move from right to left in a difference table, accumulating one level of differences to get the previous one, until you get to the output column. All we need is a method for generating the so-called *summatory polynomials*—the polynomials that give formulas for the sums of powers. And that's one of the goals of this chapter.

The simplification yields

$$1 - 5n + n^2 + 2n^3.$$

More precisely,

- In section 5.1, we'll derive formulas for the sums of powers. We'll start with a method that builds on the "hockey-stick" property of Pascal's triangle (problem 23 on page 170 of Chapter 4) and on some of the results of Chapter 1 (section 1.6) that will allow us to express our formulas in terms of the Mahler basis.
- Converting back to powers of x in these formulas will lead us to a sequence of numbers known as *Bernoulli numbers*, fascinating things in their own right. In section 5.2, we'll look more closely at these Bernoulli numbers, because they find applications all over mathematics and science. They satisfy some beautiful recurrences that allow one to quickly generate the summatory polynomials.

Around 1690, Jakob Bernoulli wrote, "With the help of [these formulas] it took me less than half of a quarter of an hour to find that the 10th powers of the first 1000 numbers being added together will yield the sum

91409924241424243424241924242500"

Pretty good for $7\frac{1}{2}$ minutes, by hand.

Bernoulli numbers were used in Andrew Wiles's famous proof in 1994 of Fermat's Last Theorem (1637).

Ways to think about it

In high school algebra, we write polynomials as linear combinations of powers of x (sums of monomials). So we are using

$$1, x, x^2, x^3, \ldots$$

as a "basis" for our polynomials. In Chapter 2, we found an interpretation (having to do with derivatives) for the coefficients: Corollary 5 on page 70 of Chapter 2 says that, if f is a polynomial of degree m,

$$f(x) = f(0) + f'(0)x + \frac{f''(0)}{2!}x^2 + \cdots + \frac{f^{(m)}(0)}{m!}x^m.$$

(*continued*)

In Chapter 1, we wrote polynomials as linear combinations of the combinatorial polynomials (the Mahler basis). So we were using

$$\binom{x}{0}, \binom{x}{1}, \binom{x}{2}, \binom{x}{3}, \dots$$

as a "basis" for our polynomials, and we found an interpretation (having to do with differences) for the coefficients: Theorem 9 on page 37 of Chapter 1 says that, if f is a polynomial of degree m,

$$f(x) = f(0) + \Delta(f)(0)\binom{x}{1} + \Delta^2(f)(0)\binom{x}{2} + \cdots + \Delta^m(f)(0)\binom{x}{m}.$$

Now we are looking for a new basis—the summatory polynomials. In a sense, we're taking as a basis for our polynomials the "formulas"

$$1, x, \frac{x(x-1)}{2}, \frac{x(x-1)(2x-1)}{6}, \dots$$

where what comes after the " ... " is one of the things we want to figure out in section 5.1. We've seen in the example above that

$$1 - 5x + x^2 + 2x^3 = 1 - 2x + 8\frac{x(x-1)}{2} + 6\frac{x(x-1)(2x-1)}{6}.$$

Maybe there's an interpretation for the "coefficients" in this new basis:

$$1, -2, 8, 6$$

Problems

Or, you can derive them now.

In these problems, feel free to use the formulas below. We'll derive them in the next section.

m	$S_m(n) = 0^m + 1^m + \cdots + (n-1)^m$
0	$n-1$
1	$\frac{n(n-1)}{2}$
2	$\frac{n(n-1)(2n-1)}{6}$
3	$\frac{n^2(n-1)^2}{4}$

1. Use the summation method to find a simple polynomial that agrees with each table.

(a)

Input	Output
0	1
1	5
2	25
3	79
4	185
5	361
6	625
7	995

(b)

Input	Output
0	−1
1	4
2	43
3	194
4	583
5	1384
6	2819
7	5158

2. Consider the result on page 202:

$$1 - 5x + x^2 + 2x^3 = 1 - 2x + 8\frac{x(x-1)}{2} + 6\frac{x(x-1)(2x-1)}{6}.$$

Study the calculation that led up to this and give an interpretation of the "coefficients"

$$1, -2, 8, 6$$

Can you find the coefficients in terms of the difference table, for example?

3. Show that if n and j are nonnegative integers,

$$\sum_{k=0}^{n-1} \binom{k}{j} = \binom{n}{j+1}.$$

Interpret this result combinatorially and in terms of Pascal's triangle.

4. One consequence of problem 3 is that

$$\binom{2}{2} + \binom{3}{2} + \binom{4}{2} + \binom{5}{2} + \binom{6}{2} + \binom{7}{2} + \cdots + \binom{n-1}{2} = \binom{n}{3}$$

Let's use this to get a formula for the sum of squares. If we replace the binomial coefficients by the corresponding rational expressions, we have

This method goes back at least to Fermat and previews the method we use in the next section.

$$\frac{2\cdot 1}{2} + \frac{3\cdot 2}{2} + \frac{4\cdot 3}{2} + \frac{5\cdot 4}{2} + \frac{6\cdot 5}{2} + \frac{7\cdot 6}{2} + \cdots + \frac{(n-1)(n-2)}{2} = \frac{n(n-1)(n-2)}{6}$$

Multiply both sides by 2 and "complicate" the left-hand side to make the form of the sum more apparent (and include a fancy 0 at the front end to make the form start at 1):

$$1(1-1) + 2(2-1) + 3(3-1) + 4(4-1) + 5(5-1) + 6(6-1) + \cdots + (n-1)(n-2) = \frac{n(n-1)(n-2)}{3}$$

We added a 0 at the front of each sum—no harm done.

Writing the last term in the sum more suggestively as $(n-1)\big((n-1)-1\big)$, the whole sum can be "split" and written as:

$$\big(0^2+1^2+2^2+3^3+\cdots+(n-1)^2\big)-\big(0+1+2+3+\cdots+(n-1)\big) = \frac{n(n-1)(n-2)}{3}$$

Now use the formula for $0+1+2+3+\cdots+(n-1)$ to isolate

$$\big(0^2+1^2+2^2+3^3+\cdots+(n-1)^2\big)$$

getting a formula for the sum of the squares from 0 to $n-1$.

5. Problem 3 implies that

$$\binom{3}{3}+\binom{4}{3}+\binom{5}{3}+\binom{6}{3}+\binom{7}{3}+\binom{8}{3}+\cdots+\binom{n-1}{3}=\binom{n}{4}$$

Starting from here, use the method of problem 4 to get a formula for

$$0^3+1^3+2^3+\cdots+(n-1)^3$$

5.1 Summatory Polynomials

The convention is to go from 0 to $n-1$ rather than 1 to n. This convention is well suited to the application we discussed in the previous section: fitting functions to tables.

The method we discuss in this section is closely tied to one used by Fermat [18]. The object of the game is to find a polynomial of smallest degree that, when evaluated at positive integers, gives you sums of powers. More precisely, let

$$S_m(n)=\sum_{k=0}^{n-1}k^m.$$

We seek a polynomial function, say $s_m(x)$ that agrees with S_m at positive integers.

Example: Suppose $m=2$. Then we are trying to find a polynomial that agrees with

$$S_2(n)=\sum_{k=0}^{n-1}k^2$$

at positive integers. That is, we want a polynomial $s_2(x)$ so that

$$s_2(n)=0^2+1^2+\cdots+(n-1)^2$$

when $n\in\mathbb{Z}^+$ $(n>0)$. We've seen such a polynomial before:

$$s_2(x)=\frac{x(x-1)(2x-1)}{6}.$$

Note that $s_2(x)$ is a cubic polynomial. And when you replace x by, say 5, you get 30, and

$$0^2+1^2+2^2+3^2+4^2=30.$$

How do we know that the $s_m(x)$ exist? That is, how do we know that there *is* a polynomial that, when evaluated at a positive integer n, gives you the sum of the mth powers between 0 and $n-1$? Well, using the case $m = 2$ as an example, look at a tabulation of S_2:

n	$S_2(n) = 0^2 + 1^2 + \cdots (n-1)^2$
1	0
2	1
3	5
4	14
5	30
6	55
7	91

As is our custom, look at the differences:

n	$S_2(n) = 0^2 + 1^2 + \cdots (n-1)^2$	$\Delta(S_2)(n)$
1	0	1
2	1	4
3	5	9
4	14	16
5	30	25
6	55	36
7	91	

The Δ column looks like perfect squares. Well, of course. If n is any positive integer,

$$\begin{aligned}\Delta(S_2)(n) &= S_2(n+1) \qquad - \quad S_2(n)\\ &= (0^2 + 1^2 + \cdots n^2) - \left(0^2 + 1^2 + \cdots (n-1)^2\right)\\ &= n^2.\end{aligned}$$

Problem 55 on page 34 of Chapter 1 says that if f is *any* function, polynomial or not, so that a polynomial of degree m agrees with $\Delta(f)$ on positive integers, then there is a polynomial of degree $m+1$ that agrees with f on positive integers.

Now invoke problem 55 on page 34 of Chapter 1 to conclude that a cubic polynomial agrees with S_2 on positive integers. The argument is the same in general (problem 6), and we have

Theorem 1. *If m is a nonnegative integer and $S_m(n)$ is defined by*

$$S_m(n) = \sum_{k=0}^{n-1} k^m,$$

then there is a polynomial $s_m(x)$ of degree $m+1$ that agrees with S_m on positive integers.

So, the s_m exist. How do we find them? Actually, we don't need any new machinery. We'll just get the Mahler expansion for $s_m(x)$ and then convert to powers of x. Let's look at an example before we try the general case.

Theorem 1 says that s_4 has degree 5, so we really only need to tabulate it between 0 and 5, but the extra numbers don't hurt anything.

Example: Let's look at the case $m = 4$: We want a formula $s_4(x)$ so that $s_4(n)$ is the sum of the fourth powers between 0 and $n - 1$. Look at a tabulation up to, say, $n = 7$.

n	$s_4(n) = S_4(n) = 0^4 + 1^4 + \cdots (n-1)^4$	$\Delta(s_4)(n) = n^4$
1	0	1
2	1	16
3	17	81
4	98	256
5	354	625
6	979	1296
7	2275	

If we have a table like this:

Input	Output	Δ
0		0
1	0	1

the missing number has to be 0. Why?

First note that if the Δ column is going to be $n \mapsto n^4$, if we added a 0th row to the table, the entry in the Δ column beside 0 would have to be 0. But that would force the entry beside 0 in the $s_2(n)$ column to be 0, too, so we can think of our table like this:

n	$s_4(n) = S_4(n) = 0^4 + 1^4 + \cdots (n-1)^4$	$\Delta(s_4)(n) = n^4$
0	0	0
1	0	1
2	1	16
3	17	81
4	98	256
5	354	625
6	979	1296
7	2275	

Now, to get the Mahler expansion for $s_4(x)$ using Newton's difference formula (Theorem 2 on page 18 of Chapter 1), we'd have to complete the difference table and look at the 0th row.

In general, "these numbers" are our friends from Chapter 4: they are the $a_{m,k}$, a.k.a. $\left\langle {m \atop k} \right\rangle$.

But wait.... Since the Δ column is given by x^4, the rest of the 0th row of the difference table is just the 0th row of the difference table for x^4. And, (see page 186 of Chapter 4), we know these numbers: they are just

$$0, 1, 14, 36, 24, 0, 0, 0, \ldots .$$

So, with no further fuss, we can complete the 0th row of the difference table (which is all we really need) for s_4:

n	$s_4(n)$	$\Delta(s_4)(n)$	Δ^2	Δ^3	Δ^4	Δ^5	Δ^6	Δ^7	$\cdots$
0	0	0	1	14	36	24	0	0	$\cdots$
1	0	1							
2	1	16							
3	17	81							
4	98	256							
5	354	625							
6	979	1296							
7	2275								

Now apply Newton's difference formula, and we have

The double bar is just to show you where we started adding columns from the difference table for x^4.

$$\begin{aligned} s_4(x) &= 0\binom{x}{0} + 0\binom{x}{1} + 1\binom{x}{2} + 14\binom{x}{3} + 36\binom{x}{4} + 24\binom{x}{5} + 0\binom{x}{6} + \cdots \\ &= 0 \cdot 1 + 0 \cdot x + 1\frac{x(x-1)}{2} + 14\frac{x(x-1)(x-2)}{6} \\ &\quad + 36\frac{x(x-1)(x-2)(x-3)}{24} + 24\frac{x(x-1)(x-2)(x-3)(x-4)}{120} \\ &= \frac{x(x-1)(2x-1)(3x^2-3x-1)}{30}. \end{aligned}$$

So, the strategy in general is to apply Newton's difference formula to the table for $s_m(x)$. But since the $\Delta(s_m)(x) = x^m$, we can get the difference table for $s_m(x)$ from the difference table for x^m by shifting it one column to the right:

We're assuming that $s_m(0) = 0$ by the same reasoning as in the previous example.

n	$s_m(n)$	$\Delta(s_m)(n)$ $= n^m$	$\Delta^2(s_m)(n)$ $= \Delta(x^m)(n)$	$\Delta^3(s_m)(n)$ $= \Delta^2(x^m)(n)$	$\Delta^4(s_m)(n)$ $= \Delta^3(x^m)(n)$	$\Delta^5(s_m)(n)$ $= \Delta^4(x^m)(n)$	$\cdots$
0							
1							
2							
3							
4							
5							
6							
7							
$\vdots$							

Furthermore, to get the Mahler expansion, all we need is the 0th row of this table. And we *know* the 0th row of the difference table for x^m—we used it extensively in Chapter 4 (and first calculated it in problem 59 on page 35 of Chapter 1). So, we have the following result:

Theorem 2. *The difference table for $s_m(x)$ has a 0th row of the form*

n	$s_m(n)$	$\Delta(s_m)(n)$	$\Delta^2(s_m)(n)$	$\Delta^3(s_m)(n)$	$\Delta^4(s_m)(n)$	$\Delta^5(s_m)(n)$	$\cdots$
0	0	$a_{m,0}$	$a_{m,1}$	$a_{m,2}$	$a_{m,3}$	$a_{m,4}$	$a_{m,5}$
1							
2							
3							
$\vdots$							

Formula 1 is from problem 59 on page 35 of Chapter 1, and formula 2 is from problem 63 on page 36 of Chapter 1.

where the $a_{m,k}$ are given by either of the following equivalent formulas:

(1) $a_{m,k} = k^m - \binom{k}{1}(k-1)^m + \binom{k}{2}(k-2)^m - \cdots \pm \binom{k}{k-1}1^m$ *or*

(2) $a_{m,k} = \begin{cases} 1 & \text{if } m = k = 0, \\ 0 & \text{if } k > m \text{ or } k < 0, \\ k(a_{m-1,k-1} + a_{m-1,k}) & \text{if } 0 \le k \le m. \end{cases}$

Applying Newton's difference formula to the result of Theorem 2, we have a wonderful theorem

Theorem 3. *If*

$$S_m(n) = \sum_{k=0}^{n-1} k^m,$$

then the unique polynomial of degree $m+1$ that agrees with S_m on positive integers is

$$\begin{aligned} s_m(x) &= a_{m,0}\binom{x}{1} + a_{m,1}\binom{x}{2} + a_{m,2}\binom{x}{3} + \cdots + a_{m,m}\binom{x}{m+1} \\ &= \sum_{k=0}^{m} a_{m,k}\binom{x}{k+1} \qquad \textit{or, in another notation} \\ &= \sum_{k=0}^{m} \left\langle {m \atop k} \right\rangle \binom{x}{k+1}. \end{aligned}$$

Remember, the $\left\langle {m \atop k} \right\rangle$ are the "lock numbers" from Chapter 4.

To see how this works, recall the table on page 186 of Chapter 4:

$k \rightarrow$ $m \downarrow$ $a_{m,k} \searrow$	0	1	2	3	4	5
0	1	0	0	0	0	0
1	0	1	0	0	0	0
2	0	1	2	0	0	0
3	0	1	6	6	0	0
4	0	1	14	36	24	0
5	0	1	30	150	240	120

This leads to summatory polynomials:

$$s_0(x) = a_{0,0}\binom{x}{1} = 1\binom{x}{1} = x,$$

$$s_1(x) = a_{1,0}\binom{x}{1} + a_{1,1}\binom{x}{2} = 0\binom{x}{1} + 1\binom{x}{2} = \frac{x(x-1)}{2},$$

$$\begin{aligned} s_2(x) &= a_{2,0}\binom{x}{1} + a_{2,1}\binom{x}{2} + a_{2,2}\binom{x}{3} = 0\binom{x}{1} + 1\binom{x}{2} + 2\binom{x}{3} \\ &= \frac{x(x-1)}{2} + 2\frac{x(x-1)(x-2)}{6} = \frac{x(x-1)(2x-1)}{6}, \end{aligned}$$

$$\begin{aligned} s_3(x) &= a_{3,0}\binom{x}{1} + a_{3,1}\binom{x}{2} + a_{3,2}\binom{x}{3} + a_{3,3}\binom{x}{4} \\ &= 0\binom{x}{1} + 1\binom{x}{2} + 6\binom{x}{3} + 6\binom{x}{4} \\ &= \frac{x(x-1)}{2} + 6\frac{x(x-1)(x-2)}{6} + 6\frac{x(x-1)(x-2)(x-3)}{24} \\ &= \frac{x^2(x-1)^2}{4}, \end{aligned}$$

Notice the shift: $a_{m,k}$ gets multiplied by $\binom{x}{k+1}$.

$$\begin{aligned} s_4(x) &= a_{4,0}\binom{x}{1} + a_{4,1}\binom{x}{2} + a_{4,2}\binom{x}{3} + a_{4,3}\binom{x}{4} + a_{4,4}\binom{x}{5} \\ &= 0\binom{x}{1} + 1\binom{x}{2} + 14\binom{x}{3} + 36\binom{x}{4} + 24\binom{x}{5} \\ &= \frac{x(x-1)}{2} + 14\frac{x(x-1)(x-2)}{6} + 36\frac{x(x-1)(x-2)(x-3)}{24} \\ &\quad + 24\frac{x(x-1)(x-2)(x-3)(x-4)}{120} \\ &= \frac{x(x-1)(2x-1)(3x^2-3x-1)}{30}. \end{aligned}$$

We'll save the calculation of the next few $s_m(x)$ for you.

We shouldn't have *all* the fun.

Ways to think about it

In principle, we can now write down the formula for the sum of mth powers. Of course, this requires calculating the $a_{m,k}$.

And then, to get the formula in terms of powers of x, you have to expand all the Mahler polynomials. But we did that generically in earlier chapters, so the process can be mechanized somewhat. Some of the problems in the next problem set will help with this.

One important feature of the method we used in this section is that it's nothing more than an application of Newton's difference formula, showing once again what a useful technique it is.

(*continued*)

A CAS can help calculate the $a_{m,k}$. In *Mathematica*, the code looks like

```
A[0, 0] := 1
 A[m_, k_] := If[Or
[k > m, k < 0], 0,
k * (A[m - 1, k - 1]+
A[m - 1, k])]
```

Finally, just as you can think of Δ as a "discrete derivative," you can think of $s_m(x)$ as a discrete *integral* of x^m. It is, after all, just a running total of the accumulated values of x^m.

Problems

6. Prove Theorem 1.

7. Here's a list of the expanded forms of the $s_m(x)$ up to $m = 4$.

m	$s_m(x)$
0	x
1	$-\frac{x}{2}+\frac{x^2}{2}$
2	$\frac{x}{6}-\frac{x^2}{2}+\frac{x^3}{3}$
3	$\frac{x^2}{4}-\frac{x^3}{2}+\frac{x^4}{4}$
4	$-\frac{x}{30}+\frac{x^3}{3}-\frac{x^4}{2}+\frac{x^5}{5}$

Extend the table to $m = 8$ and state and explain at least three patterns in the coefficients.

8. What's the sum of the first 100 squares?

9. What's the sum of the squares of the first 100 even integers?

10. What's the sum of the cubes of the first 100 odd integers?

Could you do it in less than 7.5 minutes?

11. Calculate the sum of the 10th powers of the first 1000 integers as fast as you can.

Hint: Factor out the $1/n^3$.

12. Suppose n is some fixed positive integer.

(a) Find a simple expression for

$$\frac{1}{n}\left(\left(\frac{0}{n}\right)^3+\left(\frac{1}{n}\right)^3+\left(\frac{2}{n}\right)^3+\cdots+\left(\frac{n-1}{n}\right)^3\right)$$

(b) Evaluate

$$\lim_{n\to\infty}\frac{1}{n}\left(\left(\frac{0}{n}\right)^3+\left(\frac{1}{n}\right)^3+\left(\frac{2}{n}\right)^3+\cdots+\left(\frac{n-1}{n}\right)^3\right)$$

Hint: What does $\int_0^1 x^3\,dx$ *really* mean?

(c) Where does this come up in the teaching of calculus?

Recall from Chapter 4 (see page 190, for example) that the coefficients for converting from the Mahler basis to powers of x are the $b_{m,k}$, where

$$b_{m,k} = \begin{cases} 1 & \text{if } m = k = 0 \\ 0 & \text{if } k > m \text{ or } k < 0 \\ \dfrac{b_{m-1,k-1} - (m-1)b_{m-1,k}}{m} & \text{if } 0 \le k \le m \end{cases}$$

13. By Theorem 3,

$$\begin{aligned} s_m(x) &= a_{m,0}\binom{x}{1} + a_{m,1}\binom{x}{2} + a_{m,2}\binom{x}{3} + \cdots + a_{m,m}\binom{x}{m+1} \\ &= \sum_{k=0}^{m} a_{m,k}\binom{x}{k+1} \end{aligned}$$

Replace $\binom{x}{k+1}$ by

$$b_{k+1,0} + b_{k+1,1}x + b_{k+1,2}x^2 + \cdots + b_{k+1,k+1}x^{k+1}$$

to prove Theorem 4:

Theorem 4. *An explicit formula for* $s_m(x)$ *is*

$$s_m(x) = c_{m,0} + c_{m,1}x + c_{m,2}x^2 + \cdots + c_{m,m+1}x^{m+1}$$

where

$$c_{m,j} = \sum_{k=0}^{m} a_{m,k}b_{k+1,j}$$

14. Use Theorem 4 to express $s_m(x)$ in terms of powers of x (that is, in normal form) for $m = 1, \ldots, 5$.

Check your results with problem 7.

15. Using the notation of problem 14, show that

(a) $c_{m,0} = 0$ for all $m > 0$,

(b) $c_{m,m+1} = (1/m + 1)$ for all $m \ge 0$.

So, the constant term of $s_m(x)$ is 0 and the leading coefficient is $1/m+1$.

16. In problem 58 on page 191 of Chapter 4, you showed that if we define matrices A and B by

$$A = \begin{pmatrix} a_{0,0} & a_{0,1} & a_{0,2} & \cdots & a_{0,m} \\ a_{1,0} & a_{1,1} & a_{1,2} & \cdots & a_{1,m} \\ a_{2,0} & a_{2,1} & a_{2,2} & \cdots & a_{2,m} \\ \vdots & \vdots & \vdots & \ddots & \vdots \\ a_{m,0} & a_{m,1} & a_{m,2} & \cdots & a_{m,m} \end{pmatrix}$$

and

$$B = \begin{pmatrix} b_{0,0} & b_{0,1} & b_{0,2} & \cdots & b_{0,m} \\ b_{1,0} & b_{1,1} & b_{1,2} & \cdots & b_{1,m} \\ b_{2,0} & b_{2,1} & b_{2,2} & \cdots & b_{2,m} \\ \vdots & \vdots & \vdots & \ddots & \vdots \\ b_{m,0} & b_{m,1} & b_{m,2} & \cdots & b_{m,m} \end{pmatrix}$$

then $AB = BA = I$, the identity matrix. Show that if

$$B' = \begin{pmatrix} b_{1,0} & b_{1,1} & \cdots & b_{1,m+1} \\ b_{2,0} & b_{2,1} & \cdots & b_{2,m+1} \\ b_{3,0} & b_{3,1} & \cdots & b_{3,m+1} \\ \vdots & \vdots & \ddots & \vdots \\ b_{m,0} & b_{m,1} & \cdots & b_{m,m+1} \\ b_{m+1,0} & b_{m+1,1} & \cdots & b_{m+1,m+1} \end{pmatrix}$$

then the mth row of AB' contains the $c_{m,j}$ from Theorem 4.

Ways to think about it

Think about a large matrix that contains all the $b_{m,k}$ mentioned in problem 16:

$$\begin{pmatrix} b_{0,0} & b_{0,1} & b_{0,2} & \cdots & b_{0,m} & b_{0,m+1} \\ b_{1,0} & b_{1,1} & b_{1,2} & \cdots & b_{1,m} & b_{1,m+1} \\ b_{2,0} & b_{2,1} & b_{2,2} & \cdots & b_{2,m} & b_{2,m+1} \\ b_{3,0} & b_{3,1} & b_{3,2} & \cdots & b_{3,m} & b_{3,m+1} \\ \vdots & \vdots & & \ddots & & \vdots \\ b_{m,0} & b_{m,1} & b_{m,2} & \cdots & b_{m,m} & b_{m,m+1} \\ b_{m+1,0} & b_{m+1,1} & b_{m+1,2} & \cdots & b_{m+1,m} & b_{m+1,m+1} \end{pmatrix}$$

Then matrix B can be thought of as one piece of this array:

$$\begin{pmatrix} b_{0,0} & b_{0,1} & b_{0,2} & \cdots & b_{0,m} & b_{0,m+1} \\ b_{1,0} & b_{1,1} & b_{1,2} & \cdots & b_{1,m} & b_{1,m+1} \\ b_{2,0} & b_{2,1} & b_{2,2} & \cdots & b_{2,m} & b_{2,m+1} \\ b_{3,0} & b_{3,1} & b_{3,2} & \cdots & b_{3,m} & b_{3,m+1} \\ \vdots & \vdots & & \ddots & & \vdots \\ b_{m,0} & b_{m,1} & b_{m,2} & \cdots & b_{m,m} & b_{m,m+1} \\ b_{m+1,0} & b_{m+1,1} & b_{m+1,2} & \cdots & b_{m+1,m} & b_{m+1,m+1} \end{pmatrix}$$

Matrix B

And matrix B' can be thought of as another piece:

(*continued*)

$$\begin{pmatrix} b_{0,0} & b_{0,1} & b_{0,2} & \cdots & b_{0,m} & b_{0,m+1} \\ b_{1,0} & b_{1,1} & b_{1,2} & \cdots & b_{1,m} & b_{1,m+1} \\ b_{2,0} & b_{2,1} & b_{2,2} & \cdots & b_{2,m} & b_{2,m+1} \\ b_{3,0} & b_{3,1} & b_{3,2} & \cdots & b_{3,m} & b_{3,m+1} \\ \vdots & \vdots & & \ddots & & \vdots \\ b_{m,0} & b_{m,1} & b_{m,2} & \cdots & b_{m,m} & b_{m,m+1} \\ b_{m+1,0} & b_{m+1,1} & b_{m+1,2} & \cdots & b_{m+1,m} & b_{m+1,m+1} \end{pmatrix}$$

Matrix B'

5.2 Bernoulli's Method

Here's a table of summatory polynomials up to 10:

m	$s_m(x)$
0	x
1	$-\frac{x}{2}+\frac{x^2}{2}$
2	$\frac{x}{6}-\frac{x^2}{2}+\frac{x^3}{3}$
3	$\frac{x^2}{4}-\frac{x^3}{2}+\frac{x^4}{4}$
4	$-\frac{x}{30}+\frac{x^3}{3}-\frac{x^4}{2}+\frac{x^5}{5}$
5	$-\frac{x^2}{12}+\frac{5x^4}{12}-\frac{x^5}{2}+\frac{x^6}{6}$
6	$\frac{x}{42}-\frac{x^3}{6}+\frac{x^5}{2}-\frac{x^6}{2}+\frac{x^7}{7}$
7	$\frac{x^2}{12}-\frac{7x^4}{24}+\frac{7x^6}{12}-\frac{x^7}{2}+\frac{x^8}{8}$
8	$-\frac{x}{30}+\frac{2x^3}{9}-\frac{7x^5}{15}+\frac{2x^7}{3}-\frac{x^8}{2}+\frac{x^9}{9}$
9	$\frac{3x^2}{20}+\frac{x^4}{2}-\frac{7x^6}{10}+\frac{3x^8}{4}-\frac{x^9}{2}+\frac{x^{10}}{10}$
10	$\frac{5x}{66}-\frac{x^3}{2}+x^5-x^7+\frac{5x^9}{6}-\frac{x^{10}}{2}+\frac{x^{11}}{11}$

Bernoulli must have stared very hard at this, looking for patterns. Follow in his footsteps and look for regularity before you turn the page.

Let's look at this and see what we can see. Here are some (completely unjustified) observations, based on these 11 polynomials.

- It seems that the coefficient of x^{m+1} in $s_m(x)$ is $1/m+1$.
- For $m > 0$, the coefficient of x^m in $s_m(x)$ is $-\frac{1}{2}$.

One way to try to justify these observations is to use Theorem 4 on page 211.

- For $m > 1$, the coefficient of x^{m-1} in $s_m(x)$ is $m/12$.
- For $m > 2$, the coefficient of x^{m-2} is, well, 0.
- For $m > 3$, look at the x^{m-3} term:

m	coefficient of x^{m-3} in $s_m(x)$
4	$-\frac{1}{30}$
5	$-\frac{1}{12}$
6	$-\frac{1}{6}$
7	$-\frac{7}{24}$
8	$-\frac{7}{15}$
9	$-\frac{7}{10}$
10	-1

Well, as we say in Chapter 1, let's fit a function to this table. In fact, you did that in problem 81 on page 42 of Chapter 1. You probably got

... because this is the right answer.

$$-\frac{1}{720}m(m-1)(m-2).$$

Well, we could keep going, but look at what we have so far:

$$s_m(x) = \frac{1}{m+1}x^{m+1} - \frac{1}{2}x^m + \frac{m}{12}x^{m-1} + 0x^{m-2} - \frac{1}{720}m(m-1)(m-2)x^{m-3} + \cdots .$$

So what, you say? Well, we're guessing that, after some playing around with this, Bernoulli's keen eye noticed that you could write the coefficients this way:

$$\begin{aligned} s_m(x) &= \frac{1}{m+1}\left(x^{m+1} - \frac{1}{2}(m+1)x^m + \frac{1}{12}(m+1)mx^{m-1} + 0(m+1)m(m-1)x^{m-2} \right. \\ &\quad \left. -\frac{1}{720}(m+1)m(m-1)(m-2)x^{m-3} + \cdots\right) \\ &= \frac{1}{m+1}\left(\binom{m+1}{0}x^{m+1} - \frac{1}{2}\binom{m+1}{1}x^m + \frac{1}{6}\binom{m+1}{2}x^{m-1} \right. \\ &\quad \left. +0\binom{m+1}{3}x^{m-2} - \frac{1}{30}\binom{m+1}{4}x^{m-3} + \cdots\right). \end{aligned}$$

And a conjectured pattern emerges: Perhaps the coefficient of x^{m+1-k} in $s_m(x)$ is a constant multiple of

$$\frac{1}{m+1}\binom{m+1}{k}.$$

"Finding patterns" has almost become a mantra in high school mathematics. Most often, useful patterns are *very* difficult to spot and are even more difficult to explain. And getting good at doing this requires practice.

Well, it's true. And the "constant multiples" in the conjecture have become known as *Bernoulli numbers*, denoted by B_k. We've calculated a few just now, and here are a few more:

k	0	1	2	3	4	5	6	7	8	9	10
B_k	1	$-\frac{1}{2}$	$\frac{1}{6}$	0	$-\frac{1}{30}$	0	$\frac{1}{42}$	0	$-\frac{1}{30}$	0	$\frac{5}{66}$

Let's look at one of the many ways to establish our conjecture and to generate Bernoulli's numbers.

The Details

We begin by finding a method for generating the s_m recursively. A classic argument starts with the binomial theorem (Chapter 1, page 12). Note that, for any integer i,

$$(i+1)^{m+1} - i^{m+1} = 1 + \binom{m+1}{1}i + \binom{m+1}{2}i^2 + \cdots + \binom{m+1}{m}i^m.$$

Replace i by $0, 1, 2, \ldots, n-1$:

$$\begin{aligned}
1^{m+1} - 0^{m+1} &= 1 + \tbinom{m+1}{1}0 + \tbinom{m+1}{2}0^2 + \cdots + \tbinom{m+1}{m}0^m\\
2^{m+1} - 1^{m+1} &= 1 + \tbinom{m+1}{1}1 + \tbinom{m+1}{2}1^2 + \cdots + \tbinom{m+1}{m}1^m\\
3^{m+1} - 2^{m+1} &= 1 + \tbinom{m+1}{1}2 + \tbinom{m+1}{2}2^2 + \cdots + \tbinom{m+1}{m}2^m\\
&\vdots\\
n^{m+1} - (n-1)^{m+1} &= 1 + \tbinom{m+1}{1}(n-1) + \tbinom{m+1}{2}(n-1)^2 + \cdots + \tbinom{m+1}{m}(n-1)^m.
\end{aligned}$$

Now add these equations together. On the left-hand side, everything telescopes to n^{m+1}. On the right-hand side, if you "add down," you get

$$\binom{m+1}{0}s_0(n) + \binom{m+1}{1}s_1(n) + \binom{m+1}{2}s_2(n) + \binom{m+1}{3}s_3(n) + \cdots + \binom{m+1}{m}s_m(n).$$

Since this holds for infinitely many n (namely, all positive integers), it is, by Corollary 3 on page 65 of Chapter 2, a *polynomial* identity:

$$x^{m+1} = \binom{m+1}{0}s_0(x) + \binom{m+1}{1}s_1(x) + \binom{m+1}{2}s_2(x) + \binom{m+1}{3}s_3(x) + \cdots + \binom{m+1}{m}s_m(x).$$

Imagine you had calculated $s_k(x)$ for $k = 0, 1, 2, \ldots, m-1$. You could use this identity to calculate $s_m(x)$. Indeed, since $\binom{m+1}{m} = m+1$, we can solve the equation for $s_m(x)$, proving Theorem 5:

Theorem 5. *For nonnegative integers m, we have:*

$$s_m(x) = \frac{1}{m+1}\left(x^{m+1} - \sum_{k=0}^{m-1}\binom{m+1}{k}s_k(x)\right).$$

Theorem 5 provides another way to generate summatory polynomials, one that is especially adapted to a CAS environment.

Example: Suppose you already know that

$$s_0(x) = 1, \quad s_1(x) = \frac{x(x-1)}{2}, \quad \text{and} \quad s_2(x) = \frac{x(x-1)(2x-1)}{6}.$$

Then you could use Theorem 5 to find s_3:

$$\begin{aligned} s_3(x) &= \frac{1}{4}\left(x^4 - \binom{4}{0}s_0(x) - \binom{4}{1}s_1(x) - \binom{4}{2}s_2(x)\right) \\ &= \frac{1}{4}\left(x^4 - \binom{4}{0}x - \binom{4}{1}\frac{x(x-1)}{2} - \binom{4}{2}\frac{x(x-1)(2x-1)}{6}\right) \\ &= \frac{x^2}{4} - \frac{x^3}{2} + \frac{x^4}{4}. \end{aligned}$$

Here's a *Mathematica* model of Theorem 5:

```
s[0] : = 1
s[m_] : = Expand[(1/(m + 1))*
          (x^(m + 1) - Sum[Binomial[m + 1, k]*s[k],
          {k, 0, m - 1}])]
```

The next few pages will read pretty much like a standard mathematical development. These results have been polished again and again over the years, so we thought it would be interesting to present them as a finished, very slick, product.

We can use Theorem 5 to compute the coefficients of $s_m(x)$. Remember, we suspect that the coefficient of x^{m+1-k} in $s_m(x)$ is a constant multiple of

$$\frac{1}{m+1}\binom{m+1}{k}$$

and that the "constant multiple" is what we are calling B_k. So, if $k = m$, we want the coefficient of x in $s_m(x)$ to be

$$B_m \cdot \frac{1}{m+1}\binom{m+1}{m} = B_m.$$

This leads us to *define* B_m as the coefficient of x in $s_m(x)$.

By Corollary 5 on page 70 of Chapter 2, the coefficient of x in a polynomial is the derivative of the polynomial evaluated at 0. Let's agree that $B_m = 0$ if $m < 0$.

Definition The Bernoulli number B_m is the coefficient of x in $s_m(x)$. Alternatively,

$$B_m = s_m'(0)$$

where s_m' is the derivative of s_m.

The "alternative way" in the definition of Bernoulli numbers leads to an efficient method for computing them. Here's how:

Start with

$$s_m(x) = \frac{1}{m+1}\left(x^{m+1} - \sum_{k=0}^{m-1}\binom{m+1}{k}s_k(x)\right).$$

Take the derivative of each side to get:

$$s'_m(x) = \frac{1}{m+1}\left((m+1)x^m - \sum_{k=0}^{m-1}\binom{m+1}{k}s'_k(x)\right).$$

Now substitute 0 for x to obtain a nice theorem:

Theorem 6. *The Bernoulli numbers satisfy the following recurrence:*

$$B_m = \begin{cases} 1 & \text{if } m = 0, \\ -\dfrac{1}{m+1}\left(\displaystyle\sum_{k=0}^{m-1}\binom{m+1}{k}B_k\right) & \text{if } m > 0. \end{cases}$$

Notice the similarity between this equation and the recurrence satisfied by the "lock numbers" on page 182 of Chapter 4.

Another way to state Theorem 6 is that

$$\sum_{k=0}^{m}\binom{m+1}{k}B_k = 0.$$

Example: Suppose we had completed a table of Bernoulli numbers up to $m = 5$.

k	0	1	2	3	4	5
B_k	1	$-\frac{1}{2}$	$\frac{1}{6}$	0	$-\frac{1}{30}$	0

Then we could compute B_6:

$$\begin{aligned} B_6 &= -\frac{1}{7}\left(\sum_{k=0}^{5}\binom{7}{k}B_k\right) \\ &= -\frac{1}{7}\left(\binom{7}{0}B_0 + \binom{7}{1}B_1 + \binom{7}{2}B_2\binom{7}{3}B_3 + \binom{7}{4}B_4 + \binom{7}{5}B_5\right) \\ &= -\frac{1}{7}\left(1 - \frac{7}{2} + \frac{7}{2} - \frac{7}{6}\right) = \frac{1}{42}. \end{aligned}$$

Ways to think about it

Notice what we have done: We *conjecture* that the coefficient of x^{m+1-k} in $s_m(x)$ is a constant multiple of $\frac{1}{m+1}\binom{m+1}{k}$. As we said on page 215, it's true, and the constant multiple is B_k, the kth Bernoulli number. But we're a long way from proving that. Just now, we *defined* B_m to be the coefficient of x in $s_m(x)$, so we forced our conjecture to be true for $k = m$.

(continued)

The "hope" is the same as the conjecture that the coefficient of x^{m+1-k} in $s_m(x)$ is

$$\frac{1}{m+1}\binom{m+1}{k}B_k.$$

But now we've painted ourselves into a corner: We have defined the Bernoulli numbers one way and hope that they behave in some other, more general, way. More precisely, what we hope can be stated like this:

Hope: The coefficient of x^{m+1-k} in $s_m(x)$ is

$$\frac{1}{m+1}\binom{m+1}{k} \times \text{the coefficient of } x \text{ in } s_m(x).$$

The rest of this section is dedicated to the realization of this hope.

So, the Bernoulli number B_m is the coefficient of x in the summatory polynomial $s_m(x)$. What about the other coefficients? Our suspicion is that they, too, can be expressed in terms of Bernoulli numbers. And, in fact, the method we used to get Theorem 6 can be used to get this more general result. Here's how it works:

In the sidenote above, we stated the conjecture in a slightly different way: we conjectured that the coefficient of x^{m+1-k} in $s_m(x)$ is

$$\frac{1}{m+1}\binom{m+1}{k}B_k.$$

Let $j = m+1-k$. Then $k = m+1-j$, so convince yourself that the two versions of the conjecture are the same. We'll do more of this "subscript switching" below. It's part of life.

We want to show that the coefficient of x^j in $s_m(x)$ is

$$\frac{1}{m+1}\binom{m+1}{j}B_{m+1-j}.$$

When we defined the Bernoulli numbers, we used the fact that the coefficient of x in a polynomial is the derivative of the polynomial, evaluated at 0. This followed from our work in Chapter 2. And more is true: we can describe *all* the coefficients of a polynomial in terms of the values of its derivatives at 0. Indeed, Taylor's formula (Corollary 5 on page 70 of Chapter 2) implies that

$$\begin{aligned} s_m(x) &= s_m(0) + s_m'(0)x + \frac{s_m''(0)}{2!}x^2 + \frac{s_m'''(0)}{3!}x^3 + \cdots + \frac{s_m^{(m+1)}(0)}{(m+1)!}x^{m+1} \\ &= \sum_{j=0}^{m+1} \frac{s_m^{(j)}(0)}{j!}x^j \end{aligned}$$

Notice that we already know that $s_m(0) = 0$ and $s_m'(0) = B_m$.

where $s_m^{(j)}$ is the jth derivative of s_m. So, in a sense, we "know" the coefficient of x^j in $s_m(x)$: it is

$$\frac{s_m^{(j)}(0)}{j!}.$$

But we *hope* this coefficient is a Bernoulli number times a binomial coefficient. More precisely, we hope that the following conjecture is true:

Conjecture 1. *For* $0 < j \le m+1$,

$$\frac{s_m^{(j)}(0)}{j!} = \frac{1}{m+1}\binom{m+1}{j}B_{m+1-j}.$$

You can check (problem 24) that the conjecture is true for the first few s_m, and we'll use mathematical induction to show that it's true in general. More precisely, we'll assume

The induction hypothesis (for all $k < m$):

$$\frac{s_k^{(j)}(0)}{j!} = \frac{1}{k+1}\binom{k+1}{j}B_{k+1-j} \qquad (\text{for } 0 < j \le k+1)$$

and we'll prove that the same thing holds for m.

By Theorem 5, we know that

$$s_m(x) = \frac{1}{m+1}\left(x^{m+1} - \sum_{k=0}^{m-1}\binom{m+1}{k}s_k(x)\right).$$

Differentiate this j times ($0 < j \le m+1$), replace x by 0, and divide both sides by $j!$ to get:

Notice that the jth derivative of x^{m+1} still has some "x" left, so it disappears when you replace x by 0. An exception is when $j = m+1$. We'll leave it up to you to take care of this special case.

$$\frac{s_m^{(j)}(0)}{j!} = \frac{-1}{m+1}\left(\sum_{k=0}^{m-1}\binom{m+1}{k}\frac{s_k^{(j)}(0)}{j!}\right).$$

Ah, look at the rightmost term. By the induction hypothesis, it's

$$\frac{1}{k+1}\binom{k+1}{j}B_{k+1-j}.$$

So, we have

$$\frac{s_m^{(j)}(0)}{j!} = \frac{-1}{m+1}\left(\sum_{k=0}^{m-1}\binom{m+1}{k}\frac{1}{k+1}\binom{k+1}{j}B_{k+1-j}\right).$$

What a mess. Well, let's keep going for a bit before we give up. Comparing what we have to what we want (Conjecture 1), we need to show that

$$\frac{1}{m+1}\binom{m+1}{j}B_{m+1-j} = \frac{-1}{m+1}\left(\sum_{k=0}^{m-1}\binom{m+1}{k}\frac{1}{k+1}\binom{k+1}{j}B_{k+1-j}\right).$$

Well, we can multiply by $m+1$ and divide by $\binom{m+1}{j}$. Then we are left with the job of showing that

$$B_{m+1-j} = -\frac{1}{\binom{m+1}{j}}\left(\sum_{k=0}^{m-1}\binom{m+1}{k}\frac{1}{k+1}\binom{k+1}{j}B_{k+1-j}\right).$$

The $1/\binom{m+1}{j}$ can be moved inside the sum, and we can tidy up things a bit, ending up with the following thing that we'd like to be able to prove:

$$B_{m+1-j} = -\sum_{k=0}^{m-1}\frac{\binom{m+1}{k}\binom{k+1}{j}}{\binom{m+1}{j}(k+1)}B_{k+1-j}.$$

This is a joke. We made up problem 29 on page 171 of Chapter 4 *after* we got to this point in the book and realized that this calculation would be needed here because we knew where we wanted to get (a proof of Conjecture 1). This is a (sometimes annoying) curriculum design tactic: When you need something, make it a problem in a previous chapter.

Where have we seen this coefficient of B_{k+1-j} before? Ah, yes ... it was in problem 29 on page 171 of Chapter 4, where we proved that

$$\frac{\binom{m+1}{k}\binom{k+1}{j}}{\binom{m+1}{j}} = \frac{k+1}{m+2-j}\binom{m+2-j}{k+1-j}.$$

So, what we *really* want to prove is that

$$B_{m+1-j} = -\frac{1}{m+2-j}\left(\sum_{k=0}^{m-1}\binom{m+2-j}{k+1-j}B_{k+1-j}\right). \qquad (*)$$

Oh my. This looks remarkably like Theorem 6 on page 217, which says that

$$B_m = -\frac{1}{m+1}\left(\sum_{k=0}^{m-1}\binom{m+1}{k}B_k\right).$$

Remember, $(*)$ is something we *hope* is true.

Let's write out $(*)$ to see what it really says. Suppose j is some integer between 0 and $m-1$. The front end of the sum is all zeros:

$$B_{m+1-j} = -\frac{1}{m+2-j}\Bigg(\underbrace{\binom{m+2-j}{1-j}B_{1-j} + \binom{m+2-j}{2-j}B_{2-j} + \cdots + \binom{m+2-j}{(j-1)-j}B_{-1}}_{\text{all}=0}$$

$$+\binom{m+2-j}{0}B_0 + \binom{m+2-j}{1}B_1 + \binom{m+2-j}{2}B_2 + \cdots + \binom{m+2-j}{m-j}B_{m-j}\Bigg)$$

$$= -\frac{1}{m+2-j}\left(\sum_{k=0}^{m-j}\binom{m+2-j}{k}B_k\right).$$

Let's see, what does the right-hand side of the formula in Theorem 6 say if we replace m by $m+1-j$? If m is $m+1-j$, $m-1$ is $m-j$, $m+1$ is $m+2-j$, ... oh, this is looking good. We get:

$$B_{m+1-j} = -\frac{1}{m+2-j}\left(\sum_{k=0}^{m-j}\binom{m+2-j}{k}B_k\right).$$

Bingo. This is a real fact, because it comes from Theorem 6, and it is exactly what we want—it is the re-arranged form of $(*)$.

If there were ever a time to write all this out in your own words, this is it.

So, there we have it: Conjecture 1 follows by induction, and we have a wonderful theorem, due to Bernoulli:

Theorem 7. *For* $m \geq 0$

$$s_m(x) = \frac{1}{m+1}\sum_{k=0}^{m}\binom{m+1}{k}B_k x^{m+1-k}.$$

Problems

17. Here's a list of the expanded forms of the $s_m(x)$ up to $m = 4$.

m	$s_m(x)$
0	x
1	$-\frac{x}{2} + \frac{x^2}{2}$
2	$\frac{x}{6} - \frac{x^2}{2} + \frac{x^3}{3}$
3	$\frac{x^2}{4} - \frac{x^3}{2} + \frac{x^4}{4}$
4	$-\frac{x}{30} + \frac{x^3}{3} - \frac{x^4}{2} + \frac{x^5}{5}$

Use Bernoulli's method to extend the table to $m = 8$.

18. Calculate the first 15 Bernoulli numbers. State at least two patterns in what you see.

19. Show that

$$\Delta\left(s_m'(x)\right) = m\Delta\left(s_{m-1}(x)\right)$$

20. We know that $s_m'(0) = B_m$. Investigate the numbers $s_m'(1)$ as a function of m, and find at least one conjecture.

Conjectures, anyone? Proofs?

21. We know that $s_m'(0) = B_m$. Investigate the numbers $s_m'(\frac{1}{2})$ as a function of m, and find at least one conjecture.

22. Sketch (or use a calculator to sketch) the graphs of

$$y = s_m(x) \quad \text{for } 0 \le m \le 6.$$

What points are common to more than two of the graphs? And what graphs pass through these "focal" points?

23. Let $g_m(x)$ be defined by

Compare with problem 19.

$$g_m(x) = s_m'(x) - ms_{m-1}(x)$$

Show that $\Delta(g_m) = 0$. By problem 43 on page 29 of Chapter 1, $g_m(x)$ is a constant. What is that constant?

24. Verify Conjecture 1 (page 218) for $m = 1, 2, 3, 4$, and, 5.

25. The November 1998 cover of the *Mathematics Teacher* contained a "proof", created by David Masunaga, that the sum of the cubes is the square of the sum of the integers. That is:

Notice that the sums here go from 1 to n instead of the more conventional 0 to $n-1$, but a renumbering doesn't change the main conclusion (that one sum is the square of the other).

$$\sum_{k=1}^{n} k^3 = \left(\sum_{k=1}^{n} k\right)^2$$

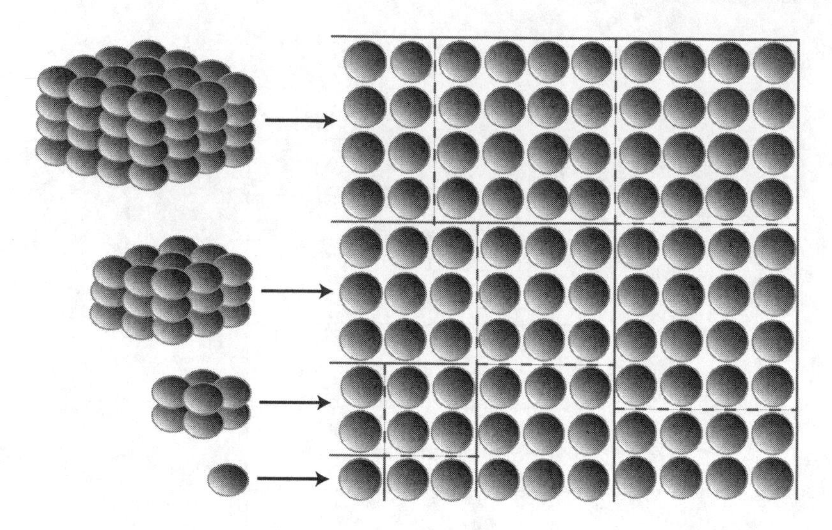

or

$$1^3 + 2^3 + 3^3 + \cdots + n^3 = (1 + 2 + 3 + \cdots + n)^2$$

Explain this proof in your own words.

26. Is the result of problem 25 a fluke? Here are the factored forms of the first 10 s_m:

(1) $\dfrac{(-1+n)n}{2}$

(2) $\dfrac{(-1+n)n(-1+2n)}{6}$

(3) $\dfrac{(-1+n)^2n^2}{4}$

(4) $\dfrac{(-1+n)n(-1+2n)\left(-1-3n+3n^2\right)}{30}$

(5) $\dfrac{(-1+n)^2n^2(-1-2n+2n^2)}{12}$

(6) $\dfrac{(-1+n)n(-1+2n)(1+3n-6n^3+3n^4)}{42}$

(7) $\dfrac{(-1+n)^2n^2(2+4n-n^2-6n^3+3n^4)}{24}$

(8) $\dfrac{(-1+n)n(-1+2n)(-3-9n-n^2+15n^3+5n^4-15n^5+5n^6)}{90}$

(9) $\dfrac{(-1+n)^2n^2(-1-n+n^2)(3+3n-n^2-4n^3+2n^4)}{20}$

(10) $\dfrac{(-1+n)n(-1+2n)(-1-n+n^2)(-5-10n+3n^2+11n^3+2n^4-9n^5+3n^6)}{66}$

State and prove at least one general result about the factored forms.

27. Suppose m is a positive integer. What happens to

$$(a+1)^m - a^m$$

if, when expanded, each power of a, a^k, is replaced by the Bernoulli number B_k?

28. Suppose m is a positive integer. Show that if you expand

$$(x+a)^{m+1} - a^{m+1}$$

and replace each power of a, a^k, by the Bernoulli number B_k, you get $(m+1)s_m(x)$.

29. Suppose f is a polynomial, say $f(x) = \sum_{i=0}^{m} c_i x^i$. State and prove a result that expresses

$$\sum_{k=0}^{n-1} f(k)$$

in terms of the coefficients of f and the summatory polynomials.

30. Suppose f is a polynomial, say $f(x) = \sum_{i=0}^{m} c_i x^i$. Show that

$$\sum_{k=0}^{n-1} f'(k)$$

can be obtained from the expansion of

$$f(x+a) - f(a)$$

by replacing x by n and each power a^k by the Bernoulli number B_k.

31. We've met, in this book, some sequences of numbers that have made appearances in several chapters and in several contexts. Among the most stellar are:

- The $a_{m,k}$ that are determined by a couple of equivalent definitions:

  (a) $a_{m,k} = k^m - \binom{k}{1}(k-1)^m + \binom{k}{2}(k-2)^m - \cdots \pm \binom{k}{k-1}1^m$

  or

  (b) $a_{m,k} = \begin{cases} 1 & \text{if } m = k = 0 \\ 0 & \text{if } k > m \text{ or } k < 0 \\ k(a_{m-1,k-1} + a_{m-1,k}) & \text{if } 0 \le k \le m \end{cases}$

- The equally lovely $b_{m,k}$ that can be defined by

$$b_{m,k} = \begin{cases} 1 & \text{if } m = k = 0 \\ 0 & \text{if } k > m \text{ or } k < 0 \\ \dfrac{b_{m-1,k-1} - (m-1)b_{m-1,k}}{m} & \text{if } 0 \le k \le m \end{cases}$$

- The famous Bernoulli numbers B_m that we've just studied and that can be generated by

$$B_m = \begin{cases} 1 & \text{if } m = 0 \\ -\dfrac{1}{m+1}\left(\displaystyle\sum_{k=0}^{m-1} \binom{m+1}{k} B_k\right) & \text{if } m > 0 \end{cases}$$

(a) Explain in a few sentences how each of these kinds of numbers has come up over the course of this book.

(b) Show that the numbers are related by the formula in Theorem 8:

Theorem 8. *For all* $m \geq 0$,

$$B_m = \sum_{k=0}^{m} a_{m,k} b_{k+1,1}.$$

32. Show that

$$B_m = \sum_{j=0}^{m} \left(\sum_{k=0}^{m} \frac{(-1)^{k+j}}{k+1} \binom{k}{j} (k-j)^m \right)$$

Problems 33–37 develop one way to prove that the odd-numbered Bernoulli numbers (after B_1) are all 0.

33. Suppose f is a polynomial. In problem 80 on page 82 of Chapter 2, you showed that

$$f(x+a+1) - f(x+a) = f'(x) + \left((a+1)^2 - a^2\right)\frac{f''(x)}{2!} + \left((a+1)^3 - a^3\right)\frac{f'''(x)}{3!} + \cdots$$

You can think of both sides of this identity as a polynomial in x whose coefficients are polynomials in a. Show that if, for all values of k, a^k is replaced with B_k, the right-hand side (and hence the left-hand side) becomes $f'(x)$.

Compare with problem 37 on page 172 of Chapter 4.

34. Show that if you replace a^k by B_k in

$$(x+a+1)^{m+1} - (x+a)^{m+1}$$

you get $(m+1)x^m$.

35. Show that if a^k is replaced by B_k in

$$a^{m+1} - (a-1)^{m+1} = \sum_{k=0}^{m} (-1)^{m-k} \binom{m+1}{k} a^k$$

you get $(m+1)(-1)^m$.

36. By problem 35,

$$(m+1)(-1)^m = (-1)^m \binom{m+1}{0} B_0 + (-1)^{m-1} \binom{m+1}{1} B_1 + \cdots + \binom{m+1}{m} B_m$$

$$= \sum_{k=0}^{m} (-1)^{m-k} \binom{m+1}{k} B_k$$

Add this to the result of Theorem 6 on page 217:

$$\sum_{k=0}^{m} \binom{m+1}{k} B_k = 0$$

to conclude that

$$\sum_{k=0}^{m} \left(1 + (-1)^{m-k}\right) \binom{m+1}{k} B_k = (m+1)(-1)^m \qquad (**)$$

37. Write out identity (**) from problem 36 without the summation signs for $m = 3, 5, 7, 9,$ and 11. Use this as inspiration to show that, for $k > 0$, $B_{2k+1} = 0$.

Problems 38–39 show another place where Bernoulli numbers appear (and another way to get results about them). They require a knowledge of the power series about the origin for the exponential function.

All you need to know is that

$$e^x = 1 + x + \frac{x^2}{2!} + \frac{x^3}{3!} + \cdots$$

38. Show that, as a formal identity,

$$\left(\sum_{n=1}^{\infty} \frac{x^n}{n!}\right)\left(\sum_{m=0}^{\infty} B_m \frac{x^m}{m!}\right) = x$$

Write both sides out without the summation signs and multiply out, "algebra style."

Use this to prove Theorem 9:

Theorem 9. *The power series (about the origin) for $x/(e^x - 1)$ is given by*

$$\frac{x}{e^x - 1} = \sum_{m=0}^{\infty} B_m \frac{x^m}{m!}.$$

39. This problem provides a method different from that of problem 37 for showing that, for $k > 0$, $B_{2k+1} = 0$.

(a) Show that

$$\frac{x}{e^x - 1} + \frac{x}{2} - 1 = \sum_{m=2}^{\infty} B_m \frac{x^m}{m!}$$

(b) If

$$f(x) = \frac{x}{e^x - 1} + \frac{x}{2} - 1$$

show that $f(-x) = f(x)$.

(c) Deduce that, for $k > 0$, $B_{2k+1} = 0$.

Facts and Notation

This is just the beginning. Bernoulli numbers show up all over mathematics. There are many known facts about these numbers—we met some of them in the problem set. You showed, for example, that after B_1, the odd-numbered Bernoulli numbers are 0. It's also known that the nonzero ones alternate in sign.

Euler showed that, for s a positive even integer, the infinite sum

$$1+\left(\frac{1}{2}\right)^s+\left(\frac{1}{3}\right)^s+\left(\frac{1}{4}\right)^s+\cdots=\sum_{n=1}^{\infty}\left(\frac{1}{n}\right)^s$$

converges to

$$\frac{(-1)^{(s/2)+1}2^{s-1}\pi^s}{s!}B_s.$$

So (when $s=2$) the sum of the reciprocals of the squares converges to $\pi^2/6$.

Fermat's Last Theorem: If $n>2$ is an integer, there are no triples of integers x, y, z with $xyz \neq 0$ and

$$x^n+y^n=z^n$$

Why is $n=2$ eliminated?

Kummer established many arithmetic properties of the Bernoulli numbers. For example, he showed that if p is a prime number and $p-1$ is not a factor of the even integer m, then the denominator of B_m is not divisible by p. He defined a "regular" prime p to be one which doesn't divide any of the numerators of the Bernoulli numbers $B_2, B_4, \ldots, B_{p-3}$. Long before Wiles's famous proof of Fermat's last theorem, it was known that, if p is a regular prime, there are no integer solutions to $x^p+y^p=z^p$ with $xyz \neq 0$. The first irregular prime is 37, and it's known that there are infinitely many irregular primes. No one knows whether or not there are infinitely many regular primes. For more on Bernoulli numbers (and a method for obtaining them that is somewhat different from the one we used in this chapter), see [9].

As you probably do with your students sometimes, we leave you a problem that we hope will challenge you, amuse you, and encourage you to do a little more reading and research.

Problem

Some of the machinery developed in the last problem set might be of use here. But the references may have some leads, too.

Show that the nonzero Bernoulli numbers alternate in sign. More precisely, show that

$$(-1)^{m+1}B_{2m}>0.$$

Notes for Selected Problems

Notes for problem 5 on page 204 Using the method from problem 4, we can show that

$$\binom{3}{3}+\binom{4}{3}+\binom{5}{3}+\cdots+\binom{n-1}{3}=\binom{n}{4}$$

can be rewritten as

$$\sum_{k=0}^{n-1}k^3-3\sum_{k=0}^{n-1}k^2+2\sum_{k=0}^{n-1}k=\frac{n(n-1)(n-2)(n-3)}{4}.$$

We'll leave it to you to fill in all the details, but here are some of the steps. After multiplying both sides by 6, the left-hand side will look like this. Note that we've added two zeros at the front:

$$1(1-1)(1-2)+2(2-1)(2-2)+3(3-1)(3-2)+\\ \cdots+(n-1)\big((n-1)-1\big)\big((n-1)-2\big)$$

To see how this sum can be split, notice that its kth term can be rewritten as

$$k(k-1)(k-2)=k^3-3k^2+2k$$

Finally, we can substitute the formulas for

$$0+1+2+\cdots+(n-1) \quad \text{and} \quad 0^2+1^2+2^2+\cdots+(n-1)^2$$

to isolate $\sum_{k=0}^{n-1} k^3$.

Notes for problem 9 on page 210 The sum that we want is

$$0^2+2^2+4^2+6^2+8^2+\cdots+198^2$$

Using the summation notation, it's

$$\sum_{k=0}^{99}(2k)^2$$

We can rewrite this sum as follows:

$$\begin{aligned}\sum_{k=0}^{99} 2^2k^2 &= 4\sum_{k=0}^{99} k^2\\ &= 4\cdot s_2(100)\\ &= 4\cdot 328350\\ &= 1313400.\end{aligned}$$

For $s_2(100)$, we used the result of problem 8.

Notes for problem 13 on page 211 Let's try an example that is general in principle. We will take $m=3$.

$$\begin{aligned}s_3(x) &= a_{3,0}\binom{x}{1}+a_{3,1}\binom{x}{2}+a_{3,2}\binom{x}{3}+a_{3,3}\binom{x}{4}\\ &= a_{3,0}(b_{1,0}+b_{1,1}x)\\ &\quad + a_{3,1}(b_{2,0}+b_{2,1}x+b_{2,2}x^2)\\ &\quad + a_{3,2}(b_{3,0}+b_{3,1}x+b_{3,2}x^2+b_{3,3}x^3)\\ &\quad + a_{3,3}(b_{4,0}+b_{4,1}x+b_{4,2}x^2+b_{4,3}x^3+b_{4,4}x^4)\end{aligned}$$

Now, note that $b_{m,k}=0$ when $k>m$. So we can rewrite the above sum by adding some extra zeros.

$$
\begin{aligned}
s_3(x) = a_{3,0}(b_{1,0} + b_{1,1}x + b_{1,2}x^2 + b_{1,3}x^3 + b_{1,4}x^4) \\
+ a_{3,1}(b_{2,0} + b_{2,1}x + b_{2,2}x^2 + b_{2,3}x^3 + b_{2,4}x^4) \\
+ a_{3,2}(b_{3,0} + b_{3,1}x + b_{3,2}x^2 + b_{3,3}x^3 + b_{3,4}x^4) \\
+ a_{3,3}(b_{4,0} + b_{4,1}x + b_{4,2}x^2 + b_{4,3}x^3 + b_{4,4}x^4)
\end{aligned}
$$

Now, we add down instead of across.

$$
\begin{aligned}
s_3(x) &= (a_{3,0}b_{1,0} + a_{3,1}b_{2,0} + a_{3,2}b_{3,0} + a_{3,3}b_{4,0}) \\
&\quad + (a_{3,0}b_{1,1} + a_{3,1}b_{2,1} + a_{3,2}b_{3,1} + a_{3,3}b_{4,1})x \\
&\quad + (a_{3,0}b_{1,2} + a_{3,1}b_{2,2} + a_{3,2}b_{3,2} + a_{3,3}b_{4,2})x^2 \\
&\quad + (a_{3,0}b_{1,3} + a_{3,1}b_{2,3} + a_{3,2}b_{3,3} + a_{3,3}b_{4,3})x^3 \\
&\quad + (a_{3,0}b_{1,4} + a_{3,1}b_{2,4} + a_{3,2}b_{3,4} + a_{3,3}b_{4,4})x^4 \\
&= c_{3,0} + c_{3,1}x + c_{3,2}x^2 + c_{3,3}x^3 + c_{3,4}x^4
\end{aligned}
$$

where $c_{3,j} = \sum_{k=0}^{3} a_{3,k}b_{k+1,j}$. Thus, in general, $c_{m,j} = \sum_{k=0}^{m} a_{m,k}b_{k+1,j}$.

Notes for problem 15a on page 211 By definition,

$$
\begin{aligned}
c_{m,0} &= \sum_{k=0}^{m} a_{m,k}b_{k+1,0} \\
&= a_{m,0}b_{1,0} + a_{m,1}b_{2,0} + \cdots + a_{m,m}b_{m+1,0}
\end{aligned}
$$

And since $b_{m,0} = 0$ for all $m > 0$, we have $c_{m,0} = 0$.

Notes for problem 15b on page 211 By definition,

$$
\begin{aligned}
c_{m,m+1} &= \sum_{k=0}^{m} a_{m,k}b_{k+1,m+1} \\
&= a_{m,0}b_{1,m+1} + a_{m,1}b_{2,m+1} + \cdots + a_{m,m-1}b_{m,m+1} + a_{m,m}b_{m+1,m+1}
\end{aligned}
$$

But note that $b_{m,k} = 0$ when $k > m$. Thus,

$$b_{1,m+1} = b_{2,m+1} = \cdots = b_{m,m+1} = 0$$

and we're left with

$$c_{m,m+1} = a_{m,m}b_{m+1,m+1} = m! \cdot \frac{1}{(m+1)!} = \frac{1}{m+1}$$

In the problems above, we used the following facts:

1. $b_{m,0} = 0$ for $m > 0$
2. $a_{m,m} = m!$
3. $b_{m,m} = 1/m!$

These can be proved using the recursion definition of $a_{m,k}$ and $b_{m,k}$. We'll leave the task to you.

Notes for problem 19 on page 221 We'll proceed by induction on m. First, we need to show that the statement is true when $m = 1$. (We'll leave this step to you.) Next, assume that

$$\Delta\big(s_k'(x)\big) = k\Delta\big(s_{k-1}(x)\big)$$

for all $k < m$. Finally, we'll show that the same holds for m.

Using Theorem 5, we have

$$s_m(x) = \frac{1}{m+1}\left(x^{m+1} - \sum_{k=0}^{m-1}\binom{m+1}{k}s_k(x)\right)$$

Differentiate both sides to get

$$s_m'(x) = x^m - \frac{1}{m+1}\sum_{k=0}^{m-1}\binom{m+1}{k}s_k'(x)$$

Now, apply the Δ operator to both sides (using linearity):

$$\Delta\big(s_m'(x)\big) = \Delta(x^m) - \frac{1}{m+1}\sum_{k=0}^{m-1}\binom{m+1}{k}\Delta\big(s_k'(x)\big)$$

The rightmost term, by induction hypothesis, is $k\Delta\big(s_{k-1}(x)\big)$. Thus, we have

$$\Delta\big(s_m'(x)\big) = \Delta(x^m) - \frac{1}{m+1}\sum_{k=0}^{m-1}\binom{m+1}{k}k\Delta\big(s_{k-1}(x)\big)$$

Now, we will need to adjust the summation a bit. Convince yourself that the following step is true.

$$\Delta\big(s_m'(x)\big) = \Delta(x^m) - \sum_{k=0}^{m-2}\binom{m+1}{k+1}\frac{k+1}{m+1}\Delta\big(s_k(x)\big)$$

And using the identity

$$\binom{m+1}{k+1}\frac{k+1}{m+1} = \binom{m}{k}$$

we get

$$\Delta\big(s_m'(x)\big) = \Delta(x^m) - \sum_{k=0}^{m-2}\binom{m}{k}\Delta\big(s_k(x)\big) \qquad (*)$$

To finish the proof, we need to show that $m\Delta\big(s_{m-1}(x)\big)$ equals the right-hand side of $(*)$. Again using Theorem 5, we get

$$\begin{aligned} m\Delta\big(s_{m-1}(x)\big) &= m\Delta\left[\frac{1}{m}\left(x^m - \sum_{k=0}^{m-2}\binom{m}{k}s_k(x)\right)\right] \\ &= \Delta(x^m) - \sum_{k=0}^{m-2}\binom{m}{k}\Delta\big(s_k(x)\big) \end{aligned}$$

Notes for problem 27 on page 222 Using the binomial theorem, we have

$$(a+1)^m = a^0 + \binom{m}{1}a^1 + \binom{m}{2}a^2 + \binom{m}{3}a^3 + \cdots + \binom{m}{m-1}a^{m-1} + a^m$$

If we subtract a^m and replace a^k by B_k, we get

$$B_0 + \binom{m}{1}B_1 + \binom{m}{2}B_2 + \binom{m}{3}B_3 + \cdots + \binom{m}{m-1}B_{m-1}$$

According to Theorem 6, this sum equals zero. We'll leave the details to you, but you need to show that

$$B_0 + \binom{m}{1}B_1 + \binom{m}{2}B_2 + \binom{m}{3}B_3 + \cdots + \binom{m}{m-2}B_{m-2} = -mB_{m-1}$$

Notes for problem 29 on page 223 Let

$$f(x) = \sum_{i=0}^{m} c_i x^i = c_0 + c_1 x + c_2 x^2 + \cdots + c_m x^m$$

Then,

$$\begin{aligned}
\sum_{k=0}^{n-1} f(k) &= \sum_{k=0}^{n-1}(c_0 + c_1 k + c_2 k^2 + \cdots + c_m k^m) \\
&= c_0 \sum_{k=0}^{n-1} 1 + c_1 \sum_{k=0}^{n-1} k + c_2 \sum_{k=0}^{n-1} k^2 + \cdots + c_m \sum_{k=0}^{n-1} k^m \\
&= c_0 S_0(n) + c_1 S_1(n) + c_2 S_2(n) + \cdots + c_m S_m(n) \\
&= \sum_{i=0}^{m} c_i S_i(n)
\end{aligned}$$

Notes for problem 30 on page 223 If we begin with $f(x+a) - f(a)$ and replace x by n, we get

$$\begin{aligned}
f(n+a) - f(a) &= \sum_{i=0}^{m} c_i (n+a)^i - \sum_{i=0}^{m} c_i a^i \\
&= \sum_{i=0}^{m} c_i \left[(n+a)^i - a^i\right] \qquad (*)
\end{aligned}$$

In $(*)$, we expand $(n+a)^i - a^i$ and replace each power a^k by B_k. According to problem 28, this gives us $i \cdot S_{i-1}(n)$. Thus, our final sum is

$$\sum_{i=0}^{m} c_i \; i \cdot S_{i-1}(n)$$

Now we must show that this sum equals

$$\sum_{k=0}^{n-1} f'(k)$$

Since

$$f'(k) = \sum_{i=0}^{m} c_i \; i \cdot k^{i-1}$$

we have

$$\begin{aligned}
\sum_{k=0}^{n-1} f'(k) &= \sum_{k=0}^{n-1} \left[\sum_{i=0}^{m} c_i \; i \cdot k^{i-1} \right] \\
&= \sum_{i=0}^{m} \left[\sum_{k=0}^{n-1} c_i \; i \cdot k^{i-1} \right] \\
&= \sum_{i=0}^{m} c_i \; i \left[\sum_{k=0}^{n-1} k^{i-1} \right] \\
&= \sum_{i=0}^{m} c_i \; i \cdot S_{i-1}(n)
\end{aligned}$$

as desired.

Notes for problem 32 on page 224 First note that

$$a_{m,k} = \sum_{j=0}^{k-1} (-1)^j \binom{k}{j} (k-j)^m$$

We can extend this sum from $j = 0$ to $j = m$. The extra terms are equal to zero since $(k - j) = 0$ when $j = k$ and $\binom{k}{j} = 0$ when $j > k$. Thus, we have

$$a_{m,k} = \sum_{j=0}^{m} (-1)^j \binom{k}{j} (k-j)^m$$

Therefore, we have

$$\begin{aligned}
B_m &= \sum_{k=0}^{m} a_{m,k} \, b_{k+1,1} \\
&= \sum_{k=0}^{m} \left(\sum_{j=0}^{m} (-1)^j \binom{k}{j} (k-j)^m \right) (-1)^k \frac{1}{k+1} \qquad (*)
\end{aligned}$$

Here, we used the fact that

$$b_{k+1,1} = (-1)^k \frac{1}{k+1}$$

which we will leave up to you to show.

Now, we can rewrite $(*)$ by bringing the term $(-1)^k(1/k + 1)$ into the inner sum (i.e., distributing). Thus, we get

$$B_m = \sum_{k=0}^{m}\left(\sum_{j=0}^{m}\frac{(-1)^{k+j}}{k+1}\binom{k}{j}(k-j)^m\right)$$
$$= \sum_{j=0}^{m}\left(\sum_{k=0}^{m}\frac{(-1)^{k+j}}{k+1}\binom{k}{j}(k-j)^m\right)$$

In the very last step, we swapped the summation. Convince yourself that this can be done. (Hint: Think about adding across versus adding down.)

Notes for problem 33 on page 224 Consider the right-hand side

$$f'(x) + \left((a+1)^2 - a^2\right)\frac{f''(x)}{2!} + \left((a+1)^3 - a^3\right)\frac{f'''(x)}{3!} + \cdots$$

As shown in problem 27, $(a+1)^m - a^m$, *when expanded* and a^k replaced with B_k, becomes zero. Thus, the only remaining term is $f'(x)$.

Notes for problem 35 on page 224 Suppose we write

$$(x+a+1)^{m+1} - (x+a)^{m+1}$$

as a polynomial in a with coefficients that are polynomials in x. Thus, we get

$$(x+a+1)^{m+1} - (x+a)^{m+1} = f_m(x)a^m + f_{m-1}(x)a^{m-1} + \cdots + f_1(x)a^1 + f_0(x)a^0 \quad (*)$$

In problem 34, we showed that if we replace a^k with B_k in $(*)$, we get $(m+1)x^m$. This means

$$f_m(x)B_m + f_{m-1}(x)B_{m-1} + \cdots + f_1(x)B_1 + f_0(x)B_0 = (m+1)x^m$$

By substituting $x = -1$ above, we get

$$f_m(-1)B_m + f_{m-1}(-1)B_{m-1} + \cdots + f_1(-1)B_1 + f_0(-1)B_0 = (m+1)(-1)^m \quad (**)$$

Finally, we must show that the left-hand side of $(**)$ can be obtained by replacing a^k with B_k in $a^{m+1} - (a-1)^{m+1}$.

Well, substituting $x = -1$ into $(*)$ gives

$$a^{m+1} - (a-1)^{m+1} = f_m(-1)a^m + f_{m-1}(-1)a^{m-1} + \cdots + f_1(-1)a^1 + f_0(-1)a^0$$

Now, if we replace a^k with B_k, the right-hand side (and hence the left-hand side) becomes

$$f_m(-1)B_m + f_{m-1}(-1)B_{m-1} + \cdots + f_1(-1)B_1 + f_0(-1)B_0$$

as desired.

Notes for problem 38 on page 225 We write out both sides of the summation and multiply them as follows.

$$\left(\sum_{n=1}^{\infty}\frac{x^n}{n!}\right)\left(\sum_{m=0}^{\infty}B_m\frac{x^m}{m!}\right) = \left(x+\frac{x^2}{2!}+\frac{x^3}{3!}+\frac{x^4}{4!}+\cdots\right)$$
$$\times\left(B_0+B_1x+B_2\frac{x^2}{2!}+B_3\frac{x^3}{3!}+\cdots\right)$$
$$= B_0x+\left(B_0\frac{1}{2!}+B_1\right)x^2+\left(B_0\frac{1}{3!}+B_1\frac{1}{2!}+\frac{B_2}{2!}\right)x^3$$
$$+\left(B_0\frac{1}{4!}+B_1\frac{1}{3!}+\frac{B_2}{2!}\frac{1}{2!}+\frac{B_3}{3!}\right)x^4+\cdots$$

Verify that the coefficients in front of x^2, x^3, x^4, and so on, are all equal to zero. To see why, let's look at the coefficient in front of x^5. It looks like this:

$$\frac{B_0}{0!}\frac{1}{5!}+\frac{B_1}{1!}\frac{1}{4!}+\frac{B_2}{2!}\frac{1}{3!}+\frac{B_3}{3!}\frac{1}{2!}+\frac{B_4}{4!}\frac{1}{1!}$$

Looks familiar? We can multiply through by $\frac{5!}{5!}$ to get

$$\frac{1}{5!}\left(B_0\binom{5}{0}+B_1\binom{5}{1}+B_2\binom{5}{2}+B_3\binom{5}{3}+B_4\binom{5}{4}\right)=\frac{1}{5!}\cdot 0=0$$

where we used Theorem 6 in the last step. Thus, all the terms in the product except B_0x vanish, and so we get

$$\left(\sum_{n=1}^{\infty}\frac{x^n}{n!}\right)\left(\sum_{m=0}^{\infty}B_m\frac{x^m}{m!}\right)=B_0x=x$$

Finally, since

$$e^x-1=\sum_{n=1}^{\infty}\frac{x^n}{n!}$$

Theorem 9 follows immediately from this identity.

Notes for problem 39c on page 225 $f(-x)=f(x)$ implies

$$\sum_{m=2}^{\infty}B_m\frac{(-x)^m}{m!}=\sum_{m=2}^{\infty}B_m\frac{x^m}{m!} \qquad (*)$$

If m is odd, then $(-x)^m=-(x^m)$. So, we can deduce from $(*)$ that

$$-\frac{B_m}{m!}x^m=\frac{B_m}{m!}x^m.$$

Therefore,

$$-\frac{B_m}{m!}=\frac{B_m}{m!} \quad\text{or}\quad B_m=0.$$

Bibliography

1. Petkovšek, M., Herbert Wilf, and Doron Zeilberger, $A = B$, A. K. Peters Ltd., Wellesley, MA, 1996.
2. Birkhoff, Garrett, and S. Mac Lane, *A Survey of Modern Algebra*, Fourth Edition, Macmillan, New York, 1977.
3. Berlinghoff, William and Fernando Gouvea, *Math Through the Ages: A Gentle History for Teachers and Others*, MAA, Washington, DC, 2003.
4. Boole, George, *A Treatise on the Calculus of Finite Differences*, Macmillan and Co., London, 1860.
5. Harvey, B., *Computer Science Logo Style*, (3 vols), MIT Press, 1997.
6. Niven, Ivan, *The Mathematics of Choice: Counting Without Counting*. MAA, Washington, DC, 1965.
7. Devaney, R., *A First Course in Chaotic Dynamical Systems Theory and Experiment*. Addison-Wesley Publishing, Reading, MA, 1992.
8. Barbeau, E.J., *Polynomials*. Springer-Verlag, New York, 1989.
9. Ireland, K. and M. Rosen, *A Classical Introduction to Modern Number Theory*, Springer-Verlag, New York, 1982.
10. Rivlin, T. J., *The Chebychev Polynomials*, Wiley, New York, 1974.
11. Graham, R.L., D.E. Knuth, and O. Patasknik, *Concrete Mathematics*, Addison-Wesley Publishing Company, Reading, MA, 1992.
12. Cuoco, A., Meta Problems in Mathematics, *College Mathematics Journal*, 31: 5 (2000), 373–378.
13. Klamkin, Murray S. On Proving Trigonometric Identities, *Mathematics Magazine*, 56:4 (September, 1983).
14. Cuoco, A., Raising the Roots, *Mathematics Magazine*, 72:5 (1999) 377–383.
15. EDC. *The CME Project*, a comprehensive high school program funded by NSF grant ESI 02 42476, forthcoming from Prentice Hall, $\approx$ 2006.
16. Antonellis, N., Dandelin Spheres. in Reader Reflections, *The Mathematics Teacher*, 96:1 (2003).

17. Needham, Tristam, *Visual Complex Analysis*, Clarendon Press, Oxford, 1997.

18. Edwards, A.W.F., *Pascal's Arithmetical Triangle*, The Johns Hopkins University Press, Baltimore, MD, 2002.

19. Benjamin, Arthur T. and Jennifer J. Quinn, *Proofs That Really Count: The Art of Combinatorial Proof*, MAA, Washington, DC, 2003.

20. Stillwell, J., *Elements of Algebra: Geometry, Numbers, Equations*, Springer-Verlag, New York, 1994.

21. Velleman, D.J. and Gregory S. Call, Permutations and Combination Locks, *Mathematics Magazine*, 68:4 (October, 1995).

Index

About the Author

Al Cuoco is Senior Scientist and Director of the Center for Mathematics Education at Education Development Center. He taught high school mathematics to a wide range of students in the Woburn, Massachusetts public schools from 1969 until 1993. At EDC, he has worked in curriculum development, professional development, and education policy. A student of Ralph Greenberg, he received a PhD from Brandeis in 1980, with a thesis and research in algebraic number theory. His favorite publication (except for this book) is his 1991 article in the *American Mathematical Monthly*, described by his wife as "an attempt to explain a number system no one understands with a picture no one can see." His new grandson, Atticus, will be talking by the time this book appears.